ENGINE BUILDER'S
HANDBOOK

HOW TO REBUILD YOUR ENGINE TO ORIGINAL OR IMPROVED CONDITION

Tom Monroe

HPBooks

HPBooks
Published by the Penguin Group
Penguin Group (USA) Inc.
375 Hudson Street, New York, New York 10014, USA
Penguin Group (Canada), 90 Eglinton Avenue East, Suite 700, Toronto, Ontario M4P 2Y3, Canada
(a division of Pearson Penguin Canada Inc.)
Penguin Books Ltd., 80 Strand, London WC2R 0RL, England
Penguin Group Ireland, 25 St. Stephen's Green, Dublin 2, Ireland (a division of Penguin Books Ltd.)
Penguin Group (Australia), 250 Camberwell Road, Camberwell, Victoria 3124, Australia
(a division of Pearson Australia Group Pty. Ltd.)
Penguin Books India Pvt. Ltd., 11 Community Centre, Panchsheel Park, New Delhi—110 017, India
Penguin Group (NZ), 67 Apollo Drive, Mairangi Bay, Auckland 1311, New Zealand
(a division of Pearson New Zealand Ltd.)
Penguin Books (South Africa) (Pty.) Ltd., 24 Sturdee Avenue, Rosebank, Johannesburg 2196,
South Africa

Penguin Books Ltd., Registered Offices: 80 Strand, London WC2R 0RL, England

While the author has made every effort to provide accurate telephone numbers and Internet addresses at the time of publication, neither the publisher nor the author assumes any responsibility for errors, or for changes that occur after publication. Further, publisher does not have any control over and does not assume any responsibility for author or third-party websites or their content.

Copyright © 1996 by Tom Monroe
Book design and production by Bird Studios
Interior photos by Tom Monroe unless otherwise noted

All rights reserved.
No part of this book may be reproduced, scanned, or distributed in any printed or electronic form without permission. Please do not participate in or encourage piracy of copyrighted materials in violation of the author's rights. Purchase only authorized editions. HPBooks is a trademark of Penguin Group (USA) Inc.

First edition: August 1996

Library of Congress Cataloging-in-Publication Data

Monroe, Tom, 1940–.
    Engine builder's handbook: inspection, machining, reconditioning, valvetrain assembly, blueprinting,
  degreeing cams, tools, engine assembly / by Tom Monroe.—1st ed.
    p. cm.
  Includes index
  ISBN 978-1-55788-245-5
    1. Automobiles—Motors—Maintenance and repair. I. Title.
TL210.M568   1996
629.25'04'0288—dc20   96-7243
                                                                           CIP

PRINTED IN THE UNITED STATES OF AMERICA

NOTICE: The information in this book is true and complete to the best of our knowledge. All recommendations on parts and procedures are made without any guarantees on the part of the author or the publisher. Tampering with, altering, modifying or removing any emissions-control device is a violation of federal law. Author and publisher disclaim all liability incurred in connection with the use of this information.

ACKNOWLEDGMENTS

Thanks to those who helped me as I photographed, wrote and did the research necessary to put together the information you see between the pages of this book. Allen Johnson allowed me to work with him while he built and tested NASCAR Grand National stock car engines, sharing information along the way. Many of the photos you see here were made possible by Morgan Shepard, who gave me access to his race engine building and testing facility. Then there was Denny Wyckoff, of Motor Machine, who shared his shop to build engines and photograph each step. And, special thanks to my wife, Patti, who encouraged me and tolerated the many hours alone while I researched, photographed and pounded away on the keyboard. ■

CONTENTS

INTRODUCTION v

CHAPTER 1
ENGINE INSPECTION 1

CHAPTER 2
ENGINE BUILDING TOOLS 17

CHAPTER 3
ENGINE TEARDOWN 29

CHAPTER 4
SHORT BLOCK 43

CHAPTER 5
CYLINDER HEADS 76

CHAPTER 6
ENGINE ASSEMBLY 108

CHAPTER 7
CAMSHAFT THEORY 154

INDEX 166

INTRODUCTION

When I was the automotive editorial director at HPBooks, I was constantly getting requests from frustrated automotive enthusiasts who needed information on rebuilding or modifying engines other than the more popular Chevy, Ford and MoPar V8s. We didn't have any. It wasn't economical for us or any publisher to go to the expense of publishing in-depth books on low-volume engines. That's why I decided to do this book, one that covers basics that apply to any engine. Clearances for main and rod bearings are basically the same; there are only so many ways a valve can be opened and closed; and air flows through a less-popular engine just as it would through one of those that's as numerous as ants. You can find most of the information you need for your engine here. I do recommend, though, you find a factory manual.

Factory manuals have the necessary specifications and illustrations that point out mechanical details of specific engines. On the other hand, they are notoriously short on information about rebuilding techniques, descriptions and tools. As an example, the term "reverse the disassembly sequence . . ." is used for most reassembly procedures. And the implication that you must have special tools to work on your engine is far from the truth, but such statements can be discouraging. Be assured, with the information in this book, good mechanical skills and a reasonably well-equipped toolbox, you should have the information necessary to do the work on your pushrod engine whether you will be doing a standard rebuild or performance modification.

Rebuilding or modifying an engine seems like an ominous task if you've never done it. And ominous it is, particularly when you consider the number of components in an engine and the decisions you must make during the course of the project. However, this difficulty can be reduced dramatically in direct proportion to the information you have. The more information you have, the easier the job. The less you have, the more difficult it is—to the point of being impossible. So gather up as much other information about your engine as possible. Talk to others who have experience with the same engine. And always work safely and use the right tool for the specific job. Take nothing for granted—check everything—and be alert. Something as simple as a head gasket installed upside down can spell the difference between a successful engine-building job and an expensive disaster. Good luck. ■

Author posing between car owner Bill Fisher (left) and good buddy crew member Gordon Downing prior to setting C-Production record at Bonneville Salt Flats. Set in 1981, author's 217.849-mph record stands as of this writing despite many assaults. No-tricks, reliable engine allowed eight five-mile wide-open throttle runs without missing a beat. Use the same approach when building your engine regardless of application and you too will have a winner.

ENGINE INSPECTION 1

Because you're reading this chapter, chances are your engine is not lying around in pieces, but is still in the engine compartment and will run—more or less. You may wonder if it really needs rebuilding. If this is your situation, read this chapter before you make the final decision. You may save a lot of time, trouble and money if you find a tune-up will cure your engine's ills. On the other hand, skip this chapter if you intend to upgrade your powerplant to improve its durability or power. Or maybe you're starting with an assortment of miscellaneous parts you've gathered over time: a bare block from here, a crank from there, special heads, forged pistons or whatever. If your situation falls under one of the later scenarios, proceed directly to Chapter 2. Otherwise, read on.

DO YOU NEED TO REBUILD?

First off, don't think your engine needs to be rebuilt just because it has accumulated a lot of miles, laps or runs down the strip. Although these are factors in the rebuild-or-not-to-rebuild decision-making process, they aren't the determining ones. How an engine was maintained and operated are more important. I've seen well-

A 409 Chevy for sale at flea market. This is a "buyer-beware" situation. Your best defense is knowledge of the specific engine and a look at the bottom end. Although specific information on rebuilding such an engine is difficult to find, using sound basics detailed here will help make your project successful.

maintained engines go three rounds of the odometer before a major rebuild was necessary. Then again, I've seen engines that didn't make it to 25,000 miles because of poor maintenance or abuse before they were in dire need of a rebuild. Engines that rarely get an oil or filter change don't last long. Heavy-duty operation can also take its toll. For instance, an engine that is used for towing or regular runs at the local drag strip will need attention sooner than your average, well-maintained grocery getter.

With all that said, performance should be the primary consideration when deciding whether or not you should rebuild your engine. It must be consuming excessive amounts of oil, oil pressure is low, fuel consumption has increased or it's noticeably down on power. If oil consumption is the problem, the cause may be a malfunctioning PCV (positive crankcase ventilation) system. This can result in pressurization of the crankcase, pushing oil out of the breather cap and seals. Low oil pressure can mean a worn oil pump or excessive bearing clearances. If power or mileage is the problem, a thorough "tune-up" may solve the problem. Carburetor and ignition servicing may be all that's needed. But with an electronically controlled engine, you may have to deal with "trouble codes." This being the case, consult an expert if you don't have the equipment or information to do the necessary troubleshooting. But if you've eliminated engine controls as the problem or you simply want to determine your engine's condition, let's look at some simple tests you can make.

LUBRICATION SYSTEM

Oil consumption and oil pressure are two of the best ways to judge engine condition. High oil consumption or low oil pressure indicates increased clearances—wear—at the bores, valves or bearings.

Oil Consumption

An engine must use some oil for lubrication. Otherwise, the pistons, piston rings and valves wouldn't receive sufficient lubrication. The question then is, how much oil should an engine use? This depends on how the engine is operated, its compression and its displacement. An engine that's under constant heavy load such as a boat engine will naturally use a little more oil. The same applies for one that's constantly at higher rpm and if it has higher compression or is turbocharged. Size must be considered, too. A 7-liter

Teardown revealed bottom end components of newly rebuilt engine were worn way beyond limits. Poor cleaning job after cylinders were honed resulted in abrasive grit being circulated through lubrication system. Noted blocked oil pickup screen.

(427 cubic inch) big-block V8 will use more oil than a 2.8-liter (171 cubic inch) V6. A quart for every 1000 miles on the V8 or 600 miles for the V6 is too much.

Check for Leakage—Before you label your engine an oil burner, check whether the oil is leaking from a gasket or seal. For example, a faulty rocker-cover gasket or rear-main seal could be the culprit. Oil spots on the driveway can mean a lot of oil is being lost on the road. Check this by looking under the car immediately after parking it. If you find dripping oil and the underside of the car is wet with oil, you may have found the problem. Suspect poor engine sealing. To zero in on the problem, steam off the underside of the car and engine or use carburetor and choke cleaner, take it for a short drive and look again. Get the car up high so you can get a good look from the underside. Don't overlook checking the positive crankcase ventilation (PCV) system. High pressure in the crankcase can cause an otherwise good seal to leak.

If you find a leak, fix it. However, if your engine isn't leaking oil, let's consider other possibilities.

Piston Rings—Unless your engine is a heavy-duty truck model, it will have three rings: two compression rings and one oil ring. The top two rings share the job of sealing the combustion chamber and some minor oil control; the bottom ring handles oil control exclusively. The oil ring scrapes off most of the oil sprayed or splashed onto the cylinder wall. Some oil is left to lubricate and seal the bore.

Although the rings may not be faulty, excess oil can get past the pistons and into the combustion chamber when more oil than intended is thrown onto the cylinder walls. This is caused by excessive bearing-to-journal clearances at the connecting rods. The looser bearing clearances allow more oil to pass between the bearings and journals. This will result in an oil-pressure drop if the additional oil flow exceeds your engine's oil-pump capacity. This condition can be detected by performing an oil-pressure check.

Just as excess oil can get up past the rings and into the combustion chamber, excess combustion gases can blow down past the compression rings into the crankcase. This is referred to as *blowby*.

Pre-1964 engines were equipped with road-draft tubes. Later engines were equipped with positive crankcase ventilation (PCV) systems to pull crankcase vapors into the intake manifold. These vapors were then drawn into the combustion chambers and burned.

If blowby is too great, the PCV system will be overtaxed. The result will be blue smoke out the exhaust pipe or breather cap, if an open PCV system is used. Here, the breather is not connected to the air cleaner—a closed PCV system has the breather connected. With the closed PCV system, excess vapors enter the air cleaner, resulting in an oily filter element. Even worse, crankcase pressures build up to the point that vapors and oil are forced out front and rear main bearing seals and gaskets that would otherwise seal the crankcase.

If the engine smokes, there will be carbon and oil deposits in the combustion chambers. The valves, spark plugs, piston tops and piston ring lands will also be coated. The inside of the intake manifold and exhaust system will have oily deposits, too.

Before condemning the rings, check the PCV system, particularly the valve. If you shake it, the valve should rattle. Also, remove the breaker and place your hand over its opening with the engine running. You should feel a distinct suction. If that's OK, check for restrictions in the PCV lines. Any restrictions in the system will cause high crankcase pressures.

Valves Guides & Seals—Valve guides have a specified clearance to the valves to allow some oil to pass for lubrication. Allowable clearance is

Broken crank was caused by loose fitting vibration damper. Front main-bearing web was broken out of block after crankshaft broke.

usually less than 0.001 inch. As the clearance grows, excess oil goes down the guides. What happens next depends on whether the valve is an intake or an exhaust. At the intake, this oil ends up in the combustion chamber. As for the exhaust valves, the oil goes directly into the exhaust ports.

Oil loss down the valve guides is also aggravated by blowby. Crankcase pressure forces more oil down the guides. Put these two conditions together and your engine has problems. Excess blowby and worn valve guides indicate an engine should be rebuilt.

Valve stem seals can be worn, cracked or missing. Any one of these conditions allows excess oil to pass down the valve guides regardless of valve stem-to-guide clearance. Replacing the valve stem seals can reduce oil consumption to an acceptable level if the rest of the engine is in good shape. Valve stem seals can be replaced without removing the cylinder heads.

Cracked Oil Passage—This is a condition where pressurized oil is forced through a crack between an oil gallery and the water jacket. Oil is then mixed with engine coolant. Once the engine is shut off, oil pressure drops to zero, but coolant temperature and pressure not only remains, it increases. This may result in coolant leaking through the crack into the oil passage.

If this condition exists, there will be an "oil slick" in the engine coolant and possibly a "milk shake" in the oil pan. Check both liquids.

Blown Intake Manifold Gasket— Unlike a cracked oil passage, there is little if any oil leakage if this condition occurs. However, it bears mention because of its serious nature.

This condition is common with some engines, particularly V-types using aluminum intake manifolds. When an engine overheats, the aluminum intake manifold grows considerably more than the cast-iron cylinder heads, resulting in a shearing action between the manifold and heads. Such relative movement tears the manifold gaskets and ruins their sealing ability.

The lubrication system is not pressurized in this area, so no oil will be lost. However, a serious coolant

Performance tuning is best done with engine under load. Chassis dyno allows this to be done in shop where all engine functions can be monitored. Exhaust probe picks up samples that are routed to gas analyzer.

leak into the lubrication system will occur at the water passages between the manifold and head/s. Such a leak will cause a rapid rise in the "oil" level in the oil pan. The "milk shake" will get very serious very fast as will be indicated on the dipstick.

LOSS OF PERFORMANCE

If fuel economy has dropped severely or power is down, there are a few checks to make. Numerous problems can cause poor performance that don't require a rebuild to correct them.

First, make sure the engine is in tune. Too much or too little fuel, or faulty ignition operation can hurt performance in a big way. Also, excessive drag from the brakes or drivetrain reduces power to the drive wheels, which also results in a huge drop in fuel mileage.

A good way to check power is to take your vehicle to a shop with a chassis dynamometer. Such a dynamometer can check power at the wheels—road horsepower. The operator can monitor the various engine components, such as the fuel system and ignition system while the car is "going down the road." You'll also be able to compare power output before and after changes are made, which leaves out the guesswork.

Have the engine tuned and checked on a dyno only if you suspect the engine is in good condition, or you have the money and want to see how much power the engine has after a rebuild. If the engine hasn't had a tune-up recently, its performance should improve from having one.

Let's take a look at other causes of performance loss before we get into engine problem diagnosis.

Non-Engine-Related Problems

Before blaming the engine, be sure the rest of the drivetrain is in good condition. Make sure the clutch is not slipping if your engine is backed up by a manual transmission. Be sure an automatic transmission shifts correctly and the torque converter stator is not freewheeling. To check that the brakes are not dragging, find a place where you can coast your vehicle down a hill in neutral. Compare this to how far a known good car or truck of the same make, model and weight will coast.

Incorrect Fuel Delivery—A restricted fuel line or filter, faulty fuel pump, dirty fuel injector nozzles, electronic sensors or bad carburetor adjustments will cause poor performance. Check for any computer trouble codes, that the carburetor is clean and float level is set correctly. Make sure the throttle plates are opening and closing all the way. Check the intake system for air leaks. These will lean out the air/fuel mixture and hurt engine performance.

A shop that has a four-gas analyzer and knows how to use it can pinpoint such problems very quickly. Such an analyzer monitors oxygen, carbon monoxide, carbon dioxide and hydrocarbons at the tailpipe.

Faulty Ignition—Check the ignition system. Make sure the engine is getting a good spark at the correct time. The wrong spark plugs, faulty wires or coil, cracked or worn distributor cap and rotor or incorrect timing can cause a huge loss in performance.

Having your engine checked on a scope where you can "see" the spark and a four-gas analyzer that will tell you what's coming out the exhaust pipe will greatly simplify this task. If your engine has vacuum and/or mechanical advance distributor—if it even has a distributor—have the advance mechanism checked. The technician should be able to do this for you. If not, remove the distributor and take it to an ignition specialty shop.

Engine-Related Problems

If you have made it this far and still haven't found a problem, check for the following: blown head gasket,

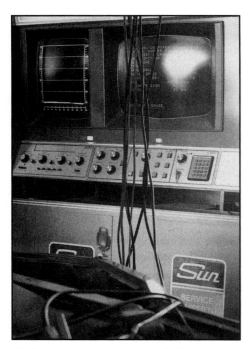

Modern scopes not only allow ignition system to be monitored, they can even check relative compression of cylinders. This tool is valuable even for troubleshooting computer-controlled engines.

restricted exhaust, worn timing chain and sprockets, worn camshaft and lifters, burned valves and carbon deposits.

Blown Head Gasket—Often there will be an immediate loss in power and an increase in engine noise with a blown head gasket. If the leak occurs to the outside of the engine you'll hear it. The noise will sound like a rapid off and on release of high pressure air. The rate increases with engine rpm.

Some head gasket leaks will enter the water jacket, showing up as multicolored oil spots in the coolant. Such a leak will also over-pressurize the cooling system, causing a rapid loss of coolant and overheating. This problem is relatively easy to diagnose. When the engine has cooled, remove the radiator cap, then start the engine. A badly blown head gasket that is leaking cylinder pressures into the water jacket will force a spout of coolant from the radiator. Another sign includes large white deposits on the spark plugs. There may also be white "smoke"—steam—from the exhaust as the engine draws coolant into the combustion chamber.

A leak into an adjacent cylinder is harder to detect. The cylinders affected should show up during a compression or leak-down test. A compression test should be done when the engine is warm because some head gaskets leak only when the engine is warm or hot. Both tests are covered beginning on page 13.

Plugged Exhaust—A restricted exhaust system has baffled many a mechanic. It didn't happen very often, but plugged exhaust systems have become more common because of double-tube exhaust pipes and damaged catalytic converters. A plugged exhaust will allow an engine to start. But the engine will either lose power or die soon afterwards depending on the degree of the restriction.

Likely causes of a plugged exhaust are a stuck heat-riser valve, a crushed pipe or muffler, a collapsed inner pipe of a double-tube exhaust or a loose baffle inside a muffler. The problem with the last two restrictions is you can't see them. This is also true in the case of a plugged catalytic converter, which occurs frequently.

There are several ways to check for a restricted exhaust system. The first is quick and easy. Simply start the engine and feel for exhaust pulses at the tailpipe. If the pulses are weak, a restriction probably exists. To check further, you'll have to raise your vehicle on a lift. With the engine running, start at the outlet end of the tailpipe and carefully touch the pipe. Do it quickly. Work forward until the pipe or muffler gets hot. Don't touch the converter. You can't touch it quick enough to avoid burns.

If the pipe is cool behind the converter, but hot ahead of it, you've found the restriction—the converter. The same goes for any other sections of the exhaust system. For example, if the pipe gets mysteriously hot, its inner tube has collapsed.

To check heat-riser valve operation of a V-type engine, wait until the exhaust cools. If your engine has one, the valve is usually located at the outlet of the right exhaust manifold. Check the shop manual for its location. The valve will either be bimetallic-spring or vacuum operated. The spring looks like a big watch spring mounted at the end of the valve shaft with a counterweight. The vacuum motor—used on 1980 and newer vehicles—has a vacuum diaphragm that operates the valve. To do this grab the counterweight or the end of the shaft and rotate it 90° to close the valve. The shaft should spring back to its resting position when you release it, thus opening the valve. If it doesn't, it can be freed with a penetrating oil. Spray the ends of the shaft and rotate it back and forth until it works smoothly so the spring or diaphragm returns it to the open position.

To test the engine to see if any portion of the exhaust system is plugged, hook up a vacuum gauge so it will read intake manifold vacuum. If vacuum is normal—over 15 inches of mercury—when the engine is cranked, but drops low or to zero once it starts and is brought to high idle, it is likely the exhaust system is plugged.

As a final check, the exhaust system can be disconnected at the exhaust manifold. Run the engine to see if the condition improves. Make sure you do this where the neighbors won't mind the noise and only long enough to make the check.

Worn Timing Set—If the timing chain and sprockets or gears are badly worn, valve timing will be off. This

5

Timing chain wear affects valve timing and ignition timing, both of which hurt engine performance. Worse, a badly worn chain and sprocket can jump timing, resulting in a no-start, no-run condition.

Burned valve cannot seal combustion chamber. Engine performance and idle quality get worse as valve continues to burn and compression drops in the affected cylinder.

can affect performance severely, even to the point of the engine not running.

A worn timing set will sometimes show up while timing the engine and working the throttle. The timing mark at the crankshaft will appear to move erratically as you slowly open and close the throttle. This can also be caused by a worn distributor, so check it also if you find this condition.

If your engine uses a cam sprocket with nylon teeth, nylon chunks in the drain oil indicate some pieces of the teeth have broken off. This means time for replacement as the teeth have become brittle.

A simple way to check timing set wear, prevalent with a chain and sprockets, is to put a socket on the crankshaft-damper bolt and use a breaker bar to turn it. Remove the distributor cap so you can see the rotor and turn the crank back and forth about 1/2 inch using the timing mark as a reference. If there is hesitation between the movement of the rotor as you change directions of crank rotation, the timing set is worn.

If the timing set is a chain and sprockets and wear is severe, the chain can jump a tooth. Generally, if the timing chain jumps one tooth the engine will barely run. If it jumps two teeth, it won't run at all. It will, however, spin fast like it has no compression, which it won't if cam timing is off by two teeth. A worn timing chain usually jumps teeth during a backfire on startup, causing camshaft timing to be advanced. It can also occur after an engine is shut off, but runs on—diesels.

Before you blame a jumped timing chain, wait a little while to make sure the hydraulic lifters haven't pumped up. This will allow the lifters to bleed down, shutting the valves and allowing the cylinders to "build" compression.

To check for a jumped timing chain, remove the top of the air cleaner. If the chain has jumped, there will usually be a steady stream of air coming out of the carburetor as the engine is cranked. Another check is to remove the distributor cap and set the crankshaft timing mark on 10° before TDC (top dead center). If the rotor doesn't point to the number-1 plug wire terminal—mark this position on the distributor housing—rotate the crankshaft 360° and check again. If it still doesn't align with the number-1 position, the chain has jumped.

Worn Cam Lobes—This problem is more prevalent in some engines, however every engine that has flat-tappet lifters—as opposed to roller lifters—has experienced the problem. If one lobe is worn more than the others, the engine will run roughly, especially under load. If all the lobes or lifters are worn excessively, engine power will be severely reduced.

If an intake valve doesn't open enough, there won't be sufficient air/fuel mixture to supply that cylinder. If an exhaust valve doesn't open enough, the incoming mixture will be diluted with exhaust gases. Either way, the engine loses power.

Burned Valves—A burned exhaust valve can't seal the combustion chamber. This results in lower compression and power. A burned valve can be caused by carbon holding the valve open or anything else that would prevent it from fully seating, retarded timing or lean air/fuel mixture. One of the major causes of burned valves is unleaded gas which lacks the heat transfer quality of the leaded variety. Sometimes the cause is a combination of more than one factor.

A burned valve can be found during

a vacuum test (p. 11), leak-down test (p. 13) or a compression test (p. 13).

Carbon Deposits—Brownish or gray/black smoke pouring from the exhaust pipe at wide open throttle is carbon being burned or knocked off.

These deposits in the combustion chambers indicate an engine that is in a poor state of tune or condition, or is not being driven properly. Carbon buildup is a result of a number of factors: rich air/fuel mixture, high oil consumption, cold operating temperature or extended low-speed driving or idling.

Detonation—Carbon deposits can cause a number of problems. First, the carbon builds up on the piston and in the cylinder head, taking up room in the combustion chamber. This raises the compression ratio. If it gets too high, the engine will detonate, or ping. During normal combustion the combustion chamber pressure builds to the point that the air/fuel mixture detonates instead of burns. This explosion can result in severe engine damage.

If the engine has a problem with detonation, don't always assume it's due to carbon buildup. Detonation can also be caused by fuel with an octane rating too low for the engine's compression ratio. Other causes may be overheating, over-advanced timing, the engine inhaling excess oil because of worn valve guides or rings, or automatic transmission fluid getting in the induction system because of a faulty vacuum modulator.

Before enough carbon accumulates to cause detonation, it usually causes preignition. The carbon particles glow hot enough to ignite the air/fuel mixture before the spark plug does—like the glow plug of a diesel or model airplane engine. This usually leads to detonation.

Both conditions are hard on the engine. They put excess stress on the pistons, rods, crank, cylinder head/s, head gasket/s and finally, your wallet.

Carbon also causes problems at the valves. As carbon builds up around a valve seat, it can prevent the valve from seating. Then the hot exhaust gases act like an acetylene torch and burn, or cut, the edge of the valve as they escape into the exhaust port. Once the edge and face is burned, the valve will not seal and the cylinder will lose compression. Meanwhile, the valve continues to burn.

Carbon deposits on valves and ports also reduce air/fuel mixture flow into the combustion chambers. At low engine speeds and during cold startup, the carbon acts like a sponge which absorbs the gasoline. The result is hard starting and low-speed driveability problems. At high engine speeds, the carbon restricts flow to the cylinder, robbing the engine of power and fuel economy.

A number of "tune-up-in-a-can" elixirs are sold to remove carbon buildup. But additives often remove carbon in big chunks. This is like throwing grit into a combustion chamber. The carbon increases ring wear and may lodge between a valve and seat.

Additives are OK if the carbon

A solid lifter valvetrain requires some lash, or clearance. This is characterized by a light clicking noise. Excess lash results in a clattering noise; too little clearance and the valve may not close fully.

buildup is light, but judging how much buildup is on the piston and cylinder head is difficult with an assembled engine. Rather than using an additive, cure the problem causing the carbon buildup. If you cure the problem, the carbon will eventually burn away.

Valve Adjustment—If the engine is equipped with solid lifters, check valve lash or clearance. A few extra thousandths of an inch of clearance between the rocker arm and valve tip can make a lot of noise and some lost power. Hydraulic lifters also make noise if the "clearance" is too great—they should have zero clearance.

A hydraulic lifter can also pump up, causing poor performance. The lifter fails to bleed down, thus preventing the valve from seating. This problem can also weaken valve springs. In either case, the engine will idle roughly, stall or not be capable of reaching high rpm.

DIAGNOSIS

With the forgoing problems in mind, you can now do the following checks to determine the condition of your engine. The most common problems have been listed, so don't

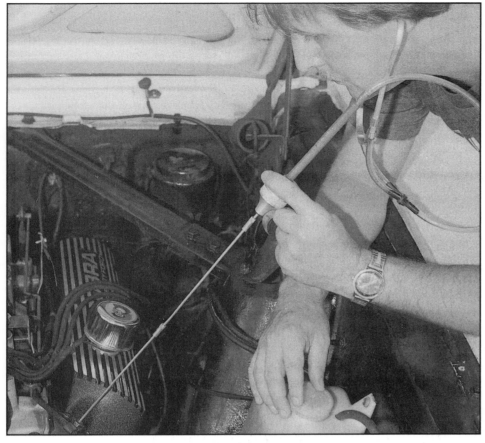

A stethoscope is very helpful when trying to pinpoint an engine noise. You must be able to distinguish between normal engine noises and those that aren't.

think if it's not mentioned, a problem doesn't exist. Yours may have an unusual problem or an entirely unique one.

Noises

Any moving part can make a noise, but finding that noise can be difficult. The first step in finding the source of the noise is eliminate the possibilities.

First determine if the noise is at engine speed or half engine speed. This will give you some clue as to where the problem is. Noises at half engine speed are usually in the valvetrain. Two exceptions are piston slap and fuel pump noises—they also occur at half speed. Noises at engine speed are normally crankshaft related, or in the bottom end of the engine.

Use a timing light to determine the speed of the noise. Hook the light to any plug wire and listen. If the noise occurs once every time the light flashes, the noise is at half engine speed. If it occurs twice for every time the timing light flashes, the noise is at engine speed.

Pinpointing Location—Once you have determined the speed of the noise, you'll need something to help with pinpointing it. A mechanic's stethoscope works best. It is flexible and picks up sound well. You can make do with a screwdriver, piece of wood or hose.

A 3-foot section of garden hose works similar to the stethoscope. Use a plastic hose if possible. It transmits sound better than rubber. A screwdriver or wooden rod transmits the sound well, but is not flexible. Flexibility is important because you'll have to get in, under and around various engine components to gain access to some areas.

Unfortunately, sounds are also carried by the cylinder block, so you may have trouble pinpointing the noise. If you use a piece of wood or a screwdriver, one end or the blade must be placed against the component you want to listen to. Press the other end against your ear or touch it to your head behind your ear.

Move around the engine until you find where the noise is loudest. Place the probe next to a spark plug. This is an excellent way to locate noise from that cylinder—especially from the piston and rod assembly.

If the noise seems to be coming from the upper end of the engine—the valvetrain—move the probe until the sound is the loudest from one of the cylinders. Alternate between intake and exhaust ports of that cylinder to pinpoint the source. Don't forget the fuel pump. It also operates at half engine speed. Read on for this problem.

Valvetrain Noise—Noise from the valvetrain is normally a clicking sound, caused by excessive clearance. If noise is coming from this area, remove the valve cover. Do a quick visual inspection. The problem may be obvious: a broken pushrod, bent pushrod, broken valve spring . . .

If all appears to be OK, check that the pushrods rotate and there's no excessive valve lash. This will have to be done with each cylinder at TDC of its firing stroke. For how to find TDC, turn to page 134. Rotate each pushrod between your fingers. If a pushrod won't rotate, lash is too tight. Next, check for excess lash by trying to rock each rocker arm back and forth. If the rocker arm "flops around," there's too much lash. Find out why.

If you still haven't found the culprit, run the engine to help locate the faulty part. If the noise occurs at low rpm, reduce idle speed so oil doesn't splash all over you and the engine compartment. Check that the pushrods rotate and the valves open and close. If a pushrod isn't rotating or a valve isn't opening as much as the rest, check more closely for a bent pushrod, or a badly worn lifter or cam lobe.

If the engine has hydraulic lifters and a noisy valvetrain, a lifter is the probable culprit. With the engine running, insert a feeler gauge between the rocker arm and valve-stem tip. If the noise quits or changes noticeably, you're in the right area. If possible, adjust the valve clearance, page 147.

If the engine has hydraulic lifters, remove the pushrod and check that it is straight, page 146. If the pushrod is OK, the problem is probably a malfunctioning lifter or it and the cam lobe is worn.

On most V-type engines, you'll have to remove the intake manifold to gain access to the lifters; on in-line engines, there will be a side cover on the block to remove. Next you must get the rocker arm out of the way. If your engine has individual rocker arm stands, loosening the rocker arms is relatively easy, page 32. For those with shaft-mounted rocker arms, you must remove the entire shaft-and-rocker arm assembly, page 32. With all that done, all you have to do now is fish the lifter out of its bore.

Piston Slap—This is caused by excessive piston-to-bore clearance. It's usually the result of piston wear or a collapsed or broken piston skirt. These are indications of high mileage, serious overheating at one time or poor cylinder-wall lubrication.

Piston slap makes a hollow or dull noise. If you were to tap a piston's thrust surface—the right side of the piston below rings—with a plastic or wood mallet, it would make a similar sound. The noise from the engine is muffled by the block and surrounding

Burned piston at top was probably caused by pre-ignition, resulting in extreme blowby, low compression and power loss. At bottom, valve head imbedded itself in piston after breaking off at high rpm. Both conditions require engine teardown to correct.

coolant.

Piston slap is loudest when an engine is cold because piston-to-bore clearance is greatest. This is particularly true with forged pistons. If the noise goes away after the engine warms up, the condition isn't serious.

To determine whether the sound you hear is piston slap, loosen the distributor hold-down bolt. With the engine running, retard ignition timing by slowly rotating the distributor in the direction of distributor shaft rotation. This will reduce the load placed on the piston, so any piston slap should become quieter or disappear altogether.

If you suspect one piston of slapping, a simple detection method is to kill one cylinder at a time while the engine is running. (This is the same method used for doing a power balance test, page 12.) This will remove all combustion load from the affected cylinder, thus quieting any piston slap.

A hole in a piston makes a noise similar to piston slap. It can be detected by removing the dipstick. You will be able to hear the noise when the piston approaches TDC and feel the resulting pulse of air from

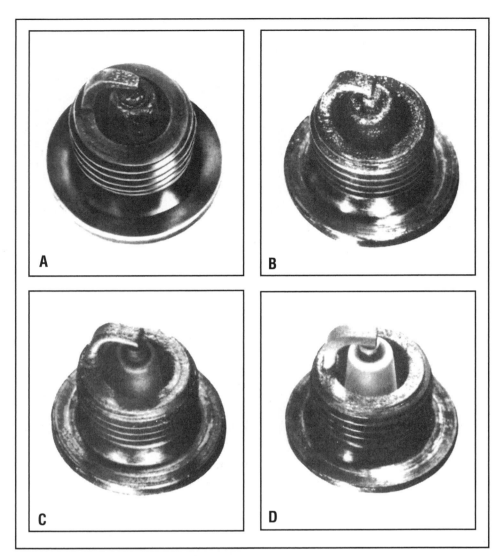

Spark plugs tell a story, so being able to read them is a good way to pinpoint a problem. Worn-out spark plugs as in A are easily recognizable by eroded electrodes and pitted insulator. New plugs will improve engine performance instantly. Oil-fouled plug B indicates internal engine wear: piston rings, cylinder bore and valve guides or seals. Carboned plug C coated with dry, black and fluffy deposits is usually from carburetion problems or driving habits. Normal spark plug D has brown to grayish tan appearance with some electrode wear indicated by slightly radiused electrode edges. When inspecting plugs keep them organized so you can isolate problem cylinders. Photos courtesy Champion Spark Plug Company.

combustion if there's a holed piston.

Main Bearing—A main bearing with excess clearance can knock at half engine speed. The noise sounds far away—muffled—in the engine. You can sometimes feel it through the accelerator pedal. The noise will usually occur when the engine is first started, hot or cold, before oil pressure builds up. It can also be detected under hard acceleration. Disabling the cylinder closest to the problem bearing journal should quiet the noise and help you to locate which one is at fault.

Don't be fooled by detonation. This noise shows up under heavy load, particularly at lower engine rpm. Detonation isn't as forceful or low pitched sounding as that caused by too much main bearing clearance. Instead, it has a tinny, rattling sound.

Connecting Rod—A connecting rod will knock or pound if oil pressure is low or its bearing-to-journal clearance is excessive. The noise will be loudest just after you let up on the throttle pedal after maintaining a constant speed.

Wrist Pin—Wrist pin noise has a double click to it and is quite pronounced. This sound is normally heard while the engine is idling or at low speeds. This problem is more likely to occur in engines with full-floating pins, rather than those with pressed pins. Incorrect clearance during assembly is the usual cause. The pin can loosen in the piston due to wear or, in the case of full-floating pins, the rod bushing can crack, flake or wear out.

Piston Rings—Faulty piston rings have a chattering sound that is most noticeable during acceleration. This is usually caused by broken rings, but it may be the result of the rings losing their tension, or springiness. This condition should show up as low compression during a compression or leak-down test.

Fuel Pump—Before you tear your engine down because of bottom end noise, check the fuel pump if it is of the mechanical variety. Fuel pump noise will normally be a fairly loud, double click at half engine speed. It's often mistaken for a bad rod.

To check the fuel pump, disconnect the fuel lines, plug them and remove the pump. Run the engine—there'll be enough fuel in the carburetor for a short period of idling. If the noise ceases, chances are the fuel pump is bad—usually a broken return spring—or the fuel pump eccentric is loose on the camshaft.

PERFORMANCE PROBLEMS

Read the Plugs

Spark plugs let you "see" what is going on inside each cylinder of an engine. Remove the plugs and keep them in order. You'll need to know which cylinder each came out of for future reference.

The two things to look for are a wet shiny black insulator or a plug with blister marks on the insulator or shell. Wet black deposits indicate excessive oil in the combustion chamber. Blistering on the insulator indicates excessive heat.

If the plug insulators have a dry gray or black coating on them, check for an over-rich mixture or weak ignition. Extensive low-speed driving or excessive idling can also cause this condition. Other problems may be a high float level, stuck choke or a sunken float.

Another possibility involves a Holley carburetor. After sitting for a long time, the gasket between the main carburetor body and metering block/s may shrink. Fuel can then run into the engine, causing a very rich mixture.

If you suspect this problem, remove the float bowl/s and metering block/s. Remove the gaskets and check them for cracks or tears. Replace them if they are bad. Keep all the gaskets soaking in gasoline while doing your check so they won't dry out. If they are good, you don't want to ruin them.

While the metering block/s is off the carburetor, check its carburetor body mounting surface. You'll need a good straightedge for making this check. This surface tends to bend or warp, preventing a good seal to the metering block regardless of gasket condition. If you think the gasket surface is uneven, take the carburetor to a Holley specialist. He'll true the mounting surface by filing or machining so the carburetor can be reused.

Vacuum Test

An internal combustion engine is basically an air pump. How well it pumps air is a good indicator of an

Vacuum gauge tells a lot about engine condition. Low vacuum—less than 15 in.Hg with street-cammed engine—usually indicates engine is in need of a rebuild. It can also be caused by retarded timing. Fluctuating needle is caused by a weak cylinder, typically from a misfiring plug, valvetrain problem or low compression in one cylinder.

engine's condition. When running, an engine produces a vacuum in the induction system between the throttle plate/s and intake valves. If each cylinder doesn't contribute its share of power, a vacuum gauge should show it by fluctuating.

Before you perform a vacuum test, make sure your engine is in good tune. A misfiring cylinder, incorrectly timed ignition and a too rich or lean air/fuel mixture will cause erratic vacuum gauge fluctuations.

The first test is done with the engine running. Hook the vacuum gauge to a fitting on the intake manifold and start the engine. At idle the reading should be about 15-22 inches of mercury (in.Hg). Altitude and cam design both affect manifold vacuum. For instance there will be 1-in.Hg less vacuum for each 1000 feet of elevation. The same goes for engines with high performance camshafts, which have increased intake-to-exhaust overlap.

A reading of 15 in.Hg or less indicates either incorrect ignition timing or a worn engine. Set the timing and recheck vacuum. If the needle floats—moves back and forth—the air/fuel mixture is probably too rich. Turn the idle mixture screws in or increase engine speed to 2000-2500 rpm to see if this corrects the problem. If the reading is low—less than 15 in.Hg—the engine may have a blown head gasket or an air leak.

Next, accelerate the engine rapidly and then release the throttle. When the engine is accelerating, the reading should drop but remain steady. If the reading fluctuates, the valve springs may be weak, allowing the valves to float—not fully seat.

When you release the throttle, the reading should jump to about 5 in.Hg above the reading at idle and then settle back to the original idle reading. If vacuum does not go that high, the pistons and rings are not sealing well.

If vacuum is normal upon startup, but soon drops to zero at idle or when you slowly increase engine speed to 2500 rpm, suspect a plugged exhaust system.

Once the engine is warm, make cranking vacuum tests. Because the engine must be cranked but not started, disable the ignition system. If the engine has a conventional point-

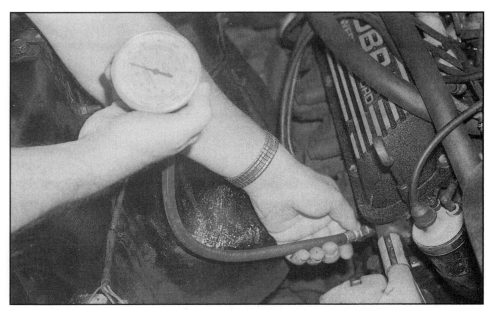

In preparation of doing compression test remove all spark plugs. Disable ignition, prop throttle and choke plates open then spin engine over four times on the compression stroke for each cylinder. Pressure of worst cylinder should be no less that 75% of best one.

type distributor, remove the points-to-coil lead and the distributor and ground it. With electronic ignitions, check the manufacturer's manual and do it to their recommendations. Electronic ignitions are easily damaged if this is done incorrectly.

Now, with either a helping hand or a remote starter, crank the engine. The vacuum gauge needle should remain fairly steady while the engine is cranked. If the needle fluctuates, one of the cylinders is not doing its share. The cause of this could be: incorrect valve adjustment, a worn camshaft lobe, collapsed lifter, a leaky valve, worn cylinder bore or piston rings, broken piston rings, a hole in the piston or a leaky head gasket. The following tests will help find which cylinder/s is at fault.

Power Balance Test

This is a good test to help you find a problem cylinder. Aircraft mechanics use a test similar to this for one of the run-up tests on reciprocating engines.

If each cylinder is contributing an equal amount of power, eliminating any one will give the same power reduction or rpm reduction. As for a weak cylinder being "killed," less power or rpm reduction will result. The opposite will result when a stronger cylinder is disabled—it will show a higher drop for each.

Most good engine analyzers with an oscilloscope have a power balance test feature. With this feature, any combination of cylinders can be disabled to perform the test.

During the power balance test, disable the cylinders one at a time while the engine is running. Note the power or rpm drop with the cylinder disabled. You'll need a writing pad and pencil to make notes as you go. Don't depend on your memory. Note the power or rpm drop and the cylinder with a rough sketch of the engine or by cylinder number. Most of us don't have a dynamometer at our disposal, so I use rpm as the reference.

For the test, idle the engine at about 1000 rpm. Disable one cylinder and allow engine speed to stabilize. Note rpm drop and the cylinder.

If your vehicle is equipped with a catalytic converter, don't allow it to run with a disabled cylinder more than five seconds. Converter damage may result from the excess fuel in the exhaust.

Caution: Be careful how you disable the cylinders. The traditional method of pulling off a spark plug lead at the plug with insulated pliers is fine with a breaker point ignition system. However, if your engine uses an electronic ignition system, particularly a distributorless ignition system (DIS), don't do it this way. You risk damaging the ignition and yourself. A jolt of electricity—up to 100,000 volts—from one of these systems has sent more than one experienced mechanic to the hospital. Don't risk it. To be absolutely certain you don't damage the ignition system or yourself, consult the manufacturer's service manual. Better yet, take your vehicle to a savvy tune-up shop where the power balance test can be done on an analyzer.

Let's look at another method of disabling each cylinder: Pull each plug wire off, then place a 4-in.-long piece of wire into the boot between the terminal and boot. Bend the wire back over the boot, then install the lead back on the plug. Do this for the remainder of the spark plugs. Bend the wire as far away from any surrounding metal as possible. By using a ground wire or jumper cable you can ground each plug individually to disable that cylinder. When doing this, don't lean on the body. This will ensure you won't provide a ground through your body.

This same method can be used at the distributor cap, but take care that there will be no arcing between the cap towers. If you use this method, keep track of which cylinder you have disabled. Use the distributor cap method if you have an engine where access to the plugs is difficult or impossible.

A final method is similar to the bent-wire approach. A number of manufacturers offer power balance test kits. It consists of a jumper wire and eight springs. The springs fit between the plug wire boots at the spark plugs or distributor cap. By touching the exposed spring with the grounded wire, each plug is shorted out with minimal risk to the ignition system or your nervous system. Just don't lean against the body sheet metal.

Compression Test

This is the easiest test you can make to determine the sealing ability of a cylinder. Compression testing should be done with the engine warm to give the most accurate indication of engine condition. Run the engine until it's up to operating temperature, then remove the plugs. As you'll soon discover, the plugs, exhaust manifold . . . everything on the engine is hot. Be careful!

Disable the ignition system. This may simply involve disconnecting the primary lead to the coil. Or if your engine has an electronic ignition, you may be able to disconnect the distributor-to-amplifier lead. Check the manufacturer's manual to be sure. Block the throttle wide open and make sure the choke is open. This will ensure the engine has unrestricted breathing during cranking.

Install the compression tester in the number-1 spark-plug hole. Crank the engine over four times—no more and no less. Record the compression pressure for this cylinder. Check the rest of the cylinders the same way and compare the readings. The lowest reading should be at least 75% of the highest for a "grocery getter." If it's a performance engine, anything below the high 90% range is unacceptable.

For example, if the highest reading is 140 psi, the lowest reading should

To isolate problem in weak cylinder, squirt a teaspoon of oil through spark plug hole and spin engine a few revolutions to distribute oil. Recheck compression. If compression improves, problem is ring-to-bore sealing; if it doesn't, problem may be a leaky valve or head gasket. Bad cam lobe or hole in piston is also a possibility.

be at least 105 psi, or 75% of 140 psi. If you want no less than a top performer, 95% of 140 psi is 133 psi, or 0.95 multiplied by 140 psi.

Caution: When doing a compression test, an explosive air/fuel mixture is being shot out of the spark plug holes when the engine is cranking. A spark or flame could cause a small explosion. So make sure there are no open flames or risk of sparks in the area where you're doing the test—and don't smoke. The hazard to your health could be immediate.

What readings should you see? The range of pressures should be 100–200 psi. They should definitely be above 100 psi, but there can be a large difference between engines. An early '70s high-performance V8 can produce up to 180 psi. If 75% is acceptable for the lowest cylinder, the lowest reading should be no less than 135 psi.

If all the cylinders except one are good do a wet compression test. Squirt about a teaspoon of motor oil through the spark plug hole. If yours is a V-type engine, direct the shot of oil toward the far side of the cylinder. Reinstall the compression gauge and check the cylinder again.

If cylinder pressure increases so it's near the others, the problem is in the bore. Chances are the rings are not sealing. If the pressure stays about the same, suspect a poor sealing valve or a blown head gasket.

Leak-Down Test

This is similar to a compression test, but with a little more sophistication, more accuracy and more expensive. The major difference between the two is that a leak-down test uses externally applied pressure. Engine cranking is not done to pressurize the cylinder being tested. The rate at which the applied pressure leaks down indicates the cylinder's sealing ability in percent leakage.

A leak-down tester won't be found in the typical toolbox. But except for the percent leakage gauge, the test can be done with a simple air tank and a plug adapter.

The leak-down tester requires a compressed-air source. If you already

> ### VALVE JOBS & HIGH MILEAGE ENGINES
>
> A valve job done on a high-mileage engine usually makes performance and oil consumption problems worse. This is because all of the parts of an engine wear together. As the sealing of the valves deteriorate, so does ring and piston sealing. Consequently, if the cylinder heads are reconditioned they seal better, resulting in higher pressure and vacuum loads on the rings and pistons. Where the rings and pistons were doing an adequate sealing job before, they may not be up to the job now, resulting in increased oil consumption and blowby. So avoid the valve-job-only solution.

have this, you can perform all the tests without the leak detector—except for determining percent loss through leakage.

A leak-down test is performed with the piston at top dead center (TDC) on its compression stroke—both valves are closed. Air is supplied to the cylinder through a special fitting that adapts the air hose to the spark plug hole. The cylinder should do a good job of holding the pressure. If it doesn't, the test will allow you to pinpoint the problem.

Caution: Be sure your hands are clear of the fan, belts or pulleys while doing a leak-down test. As the cylinder is pressurized, the engine will turn over with a short burst if the piston is not exactly on TDC.

While testing, have the breather and radiator cap off and throttle plates and choke blocked wide open. This way, you can listen for any leaks.

To set each piston at TDC, mark the distributor housing directly in line with each spark plug tower. Remove the distributor cap and rotate the crankshaft until the rotor aligns with the corresponding mark for cylinder number 1 and note the position of the rotor to cylinder-1 firing position. If the rotor doesn't align, rotate the crankshaft one turn. (Use a socket and breaker bar on the damper bolt to turn the crank.) When you're finished with checking one cylinder, turn the crankshaft until the distributor rotor aligns with the next position in the firing order and check that cylinder. Yes. Do the leak-down test in the firing order for convenience sake.

Connect the air hose and pressurize the cylinder. This is when the leak-down gauge comes in handy. The gauge tells you what percentage of air supplied to the cylinder leaks out. A small amount of air leakage through the breather is common on a worn engine. Acceptable is 5-10%; over 20% is poor.

If you don't have a leak-down gauge, listen for air leaking at the carburetor, exhaust pipe(s), radiator and oil filler, breather and dipstick tube.

If you hear air escaping through the carburetor, the intake valve is leaking. If you hear it at the tailpipe, the exhaust valve is leaking. If both valves are leaking, check crankshaft position. You may have it set on TDC between the intake and exhaust strokes.

A head gasket leak will show up as air leaking out of an adjacent cylinder or through the radiator filler neck. A cracked cylinder wall or head will also show up as air leaking from the filler neck. Look for bubbles in the coolant. If you hear leakage at the breather cap or dipstick tube, the pistons and rings are not sealing in the cylinder bore.

Head Gasket—If pressure is down on two adjacent cylinders, chances are the head gasket has blown between them. A compression test of the two cylinders will show low pressure, although not always equally. During a leak-down test you should hear air escaping from one spark plug hole while the adjacent one is pressurized and vice versa.

A head gasket can also blow between a cylinder and a water jacket. This will let combustion pressure escape into the cooling system and coolant into the combustion chamber. If the problem is severe, you can confirm the leak by pressurizing the cooling system. You'll be able to hear air or coolant escaping into that cylinder through the spark plug hole.

A blown head gasket is also indicated by the spark plugs. If one or two plugs show white fluffy deposits and the rest look OK, suspect a blown head gasket.

Valves—As I noted earlier, a leak-down test will expose a bad valve by allowing air to escape from the carburetor or injector throttle body, or the tailpipe. If you don't have a leak-down tester, you'll have to depend on a compression check and further testing. If oil didn't correct a low-

> ### BE CAREFUL
>
> Use caution when working on a running engine. Make sure you or the equipment and tools you're working with don't get caught in the fan or accessory drive pulleys and belts. Keep neckties and jewelry in your dresser drawer! And stay out of the plane of the fan. Fan blades have been known to break off and impale themselves in the bodies of mechanics like a dagger. Wear safety glasses. Be safety conscious—accidents happen.

Depth gauge end of vernier caliper is used to measure valve lift at spring retainer. Difference in measurement between valve fully closed and opened is valve lift. Similar measurement at pushrod is lobe lift.

compression reading, chances are the valves are not sealing.

The first thing to do is to make sure the valve is closing all the way. Remove the valve cover. Rotate the crankshaft until the piston for the bad cylinder is at TDC on the compression stroke—the distributor rotor will be pointing to the position where that cylinder's distributor cap tower would be. With the piston in this position you should be able to rotate both pushrods easily with your fingers. They shouldn't be loose.

If there is too much clearance—the pushrod is loose—an easy check is to remove the rocker-arm assembly. Refer to page 32 for how to do this. Place a straightedge across all the valve stem tips or valve spring retainers and see if any valves are at a different height.

There will be some variance, but it should be less than 0.015 in. from one valve to another. A valve "taller" than the others may be burned or on a sunken valve seat. A "short" one is probably being held open by a bent stem or has carbon between its face and seat.

CAMSHAFT LOBE WEAR

Cam lobe wear is a common problem in some engines and not others. If a valve is not opening all the way and hydraulic lifters are used, suspect a lifter as the culprit. If it's not the lifter, then it could be a lobe and lifter—they wear together. Such a problem occurs on high mileage engines, those with extremely stiff valve springs and new engines that aren't run in correctly.

Checking Valve Lift

Cam lobe wear shows up directly as reduced valve lift. If the engine has solid lifters, it's a simple matter to measure valve lift using the following method. If solid, or mechanical, lifters are used, proceed directly to *Checking Lobe Lift*. The pressure of the valve springs cause hydraulic lifters to bleed down—particularly on high-mileage engines. This makes it difficult to get an accurate valve lift reading.

If you have a vernier caliper or dial indicator, use it. If not, measure as closely as you can with a divider and ruler or just a ruler. To measure valve lift you must start with each lifter on the base of its cam lobe. To be sure the cam lobe is in this position just set the crank to TDC on the compression stroke of the particular cylinder.

The best place to measure with a vernier caliper is from the machined spring seat, but this may not be possible with some engines because of the way the seat is machined. Any surface will do, though, just as long as you measure up from the same point on the head for opened and closed valve positions. The difference between these two readings is valve lift.

Record the distance from the head to the top of the retainer with the valve closed. Turn the crankshaft until the valve is fully open—the spring is most compressed. Subtract the difference between the two measurements to find actual valve lift.

If you are using a dial indicator, turn the crankshaft until the valve you're going to check appears to be fully open. Firmly mount the indicator with its stand so the indicator tip is against the valve retainer and the plunger is in line—parallel—with the valve stem. Make sure the plunger is compressed a little, but allow for enough travel to read specified valve opening. Rotate the crankshaft back and forth to confirm the valve is fully seated—indicator reading will be lowest—and zero the dial. Now you're ready to check valve lift. Rotate the crankshaft until indicator reading is maximum and record the reading for the lift of that valve.

Compare readings after you've checked the last cylinder, then compare these lifts to specifications for your engine's cam. Recheck any valves that are not to specification. To get approximate lobe lift, divide valve lift by the rocker arm ratio. You'll need the specifications of your engine for this.

Lobe lift is measured directly with dial-indicator plunger tip positioned against pushrod tip. Zero indicator when lifter is on lobe base circle, then rotate engine slowly until maximum reading is obtained. This figure is lobe lift.

Checking Lobe Lift

To check the camshaft lobes directly, remove the rocker arms. Leave the pushrods in place. Measure lobe lift at the rocker-arm end of each pushrod. This is just about the most accurate way of measuring lobe lift.

Some pushrods are guided by holes in the cylinder head or intake manifold, a few by guide plates and others by the rocker arms themselves. Regardless of how they are guided, the upper end of the pushrods will be free to move in at least the plane described by the pushrod and valve. This will allow them to fall away from the dial-indicator plunger if yours is a V-type engine. To keep this from happening, position something in the head for the pushrod to rest against so it's centered over the lifter. A small deep socket or screwdriver laid down in the head should do the trick.

Another trick to use is a pushrod with a cup-type end instead of the ball type. If yours doesn't use such a pushrod, I'm sure your local engine builder will have an old one he'll let you have. It should be at least as long as the pushrods from your engine. An older Chevy or Ford six or a Y-block Ford use these type pushrods as do most engines with solid lifters with shaft-type rocker arms. Longer is OK. Just move it from one lifter to the other as you check the lobes.

With the pushrod in place, install the indicator with its base so the plunger is directly in line with the pushrod and the plunger is on the end of the pushrod. Rotate the crank until you find the lowest indicator ready, zero the indicator and rotate the crank until you get the highest reading. This last reading is actual lobe lift. Compare this to the specs for your engine. If your engine is in street trim, lift should be about 1/4 inch give or take some depending on your engine and the type of camshaft used. It could go up as high as 0.300 in. For a high-performance cam, lobe lift could approach 1/2 inch. Regardless of the type of cam, the difference between lobe lifts should be within 0.005 in.

Pay particular attention to any pushrods that didn't rotate when the engine was running. The taper on the lobe is probably worn away, which is the first sign of a bad lobe and lifter.

Also note: If any of the lobes don't measure up to specification, but the engine has the desired compression and oil consumption is OK, you can just replace the camshaft and all lifters. Never use old lifters on a new camshaft. The old lifters will quickly destroy the new cam lobes. ■

ENGINE BUILDING TOOLS 2

Other than the basic tools and equipment used in tearing down and assembling an engine, I discuss in this chapter special engine building tools and how to use them, such as those for blueprinting, inspection and assembly. You should already have the basic tools such as wrenches, a socket set, feeler gauges and screwdrivers. Even though you'll need some special tools and equipment—more if your engine building project includes precision measuring for high-performance modifications—you won't need much more than the basic tools.

To do the best job possible, you'll need to stock up on information such as this and other publications, particularly those that deal specifically with your engine. Read and ask questions. And don't skim over the basics. The best engine builders in the world stick to basics and they can't ever seem to get enough information regardless of how minor it may be. They constantly read and ask questions. You do the same.

As you progress through your engine project, you'll quickly discover that it's made up of a series of interrelated projects, projects that come together in one big project. So as you undertake each one, always plan what you're about to do

Engine building projects require a minimum of measuring and assembly tools. Consider the torque wrench a tool that falls in the must-have category. The one I'm using here is an inexpensive but accurate needle-type torque wrench. Photo by Bill Keller.

before you do it—and keep things simple.

There are three basic types of engine building tools; first are those for doing disassembly and assembly work, second are measuring tools and third are tools for preparing parts. Of the three, you're most likely to have the ones for disassembly and assembly work. Because of their special nature though, you may have few if any inspection and preparation tools, particularly those for doing blueprinting and high performance work. Let's take a look at some engine building tools.

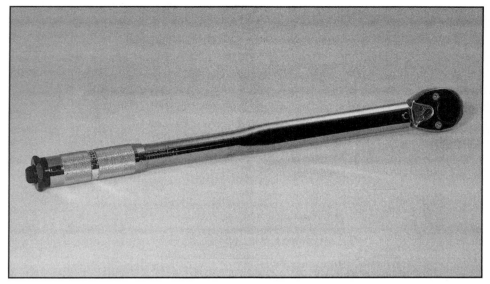

Main advantage of clicker-style torque wrench over needle or dial torque wrench is you don't have to look at it while torquing bolts and nuts.

Installed between breaker bar and socket, when preset torque of sensor is reached it buzzes. If you must have all the "bells and whistles," torque sensor is available from Snap-on.

ASSEMBLY & DISASSEMBLY TOOLS

Torque Wrench

Although a relatively common assembly tool, I mention the torque wrench because it is crucial that you use one when assembling your engine. Also crucial is knowing how to use one. I'll also dispel a few misconceptions about torque wrenches and their use.

If you don't have a torque wrench in your toolbox, get one before you begin your engine project. Correctly torqued fasteners are critical to any mechanical assembly, particularly an engine. Why torque a bolt in the first place? The primary reason is that it will not be stressed—or loaded—in service beyond the loading it receives from its initial torquing. This is critical for fasteners retaining engine components such as rod caps, main bearing caps, cylinder heads and intake manifolds. With that said let's take a look at what kind of torque wrench to get. There are two basic types: the bending beam type and the click type.

Of the two torque wrenches, you'll be pleased to know that the least expensive is the most durable. In terms of accuracy it is also the most dependable. But wouldn't you know that there's a catch? The beam torque wrench is also the most difficult to use. As for the click-type torque wrench, it requires periodic calibrating by the manufacturer to ensure it is accurate. But because it's easier to use and because it looks cooler to have one, you may be willing to stretch your budget to have a clicker torque wrench grace your toolbox.

Torquing Technique—The following rules apply to all types of torque wrenches. Don't jerk a torque wrench when tightening a fastener. Instead, use a slow steady pull on the handle with one hand while steadying the wrench at the drive end with the other. To dispel a myth, an extension between a torque wrench and socket has no effect on torque as would be the case when used with an impact wrench. Following are some specifics concerning the two types of torque wrenches.

Before you use a beam torque wrench, always check that the pointer aligns with 0 on the scale. If it doesn't, bend the needle so it does align. Additionally, make sure the scale doesn't touch the slot in the guide or drag on the scale. Both cause inaccurate readings. As you torque a bolt, pull so the handle is balanced on its pivot and look squarely at the scale, not at an angle. The need to look at the scale with your face perpendicular to it is what makes a beam-type torque wrench more difficult to use than the click type. Torquing bolts in tight areas such as those for cylinder heads on an installed engine can present a problem. Assembling an engine on a stand or bench shouldn't be. Correct torque is reached when the needle aligns with the desired value on the scale.

To use a click-type torque wrench, unlock the handle or knob and rotate it until the index mark at its edge of the handle or scale in the window aligns with the desired torque. Lock the handle in this position. As you tighten a bolt or nut, the torque you set is signaled with an audible click. Stop pulling when you hear the click.

Dial Torque Wrench—There's a third style of torque wrench that isn't used much because it incorporates many of the undesirable features of both the beam and click-type torque wrenches. The dial torque wrench is expensive, needs periodic calibration and has a scale you must watch as you torque a bolt. I don't recommend this type of torque wrench, but if you are "well healed," these are available with a device that emits an audible beep when the preset torque is reached.

Some final words about torque wrenches: Regardless of the type you choose, get one that reads in foot-pounds/Newton-meters (ft-lbs./Nm) not in inch-pounds (in-lbs.) and has a sufficient range, but not one that's too broad. Generally, the broader the torque range, the more expensive the wrench. With all things being equal, accuracy is not as good with a broader range. For example, the typical beam wrench has a 0-100 ft-lb./0-138Nm range and will be accurate across the total range. Another example is a 50-250 ft-lb./70-345Nm click-type torque wrench. It's much more expensive, but its accuracy is not as good. A wrench with a 50-150 ft-lb. range would be better. As for drive size, get a 1/2-inch-drive torque wrench, not a 3/8-inch-drive. Get a 1/2-to-3/8 adapter, though. Chances are you'll be using 3/8-in. drive sockets more than 1/2-in. drive sockets.

Angle Gauge

Although not many pushrod engines use bolts that must be rotated through a specific angle after they are torqued, all auto manufacturers now use them. Many of these bolts are the *torque-to-yield* type. They are permanently stretched during the tightening process and cannot be reused.

Because of the increased use of this and other tightening procedures, you should be aware of the angle gauge. This tool installs between the wrench and socket. For example, if a bolt is to be torqued 45 ft-lbs. and rotated an additional 60°, simply torque the bolt to 45 ft-lbs., zero the angle gauge, then rotate the bolt another 60° as indicated by the gauge.

Check tightening specs and recommendations for your engine to determine if you need such a tool. Because angle torquing is becoming more prevalent, angle gauges are readily available and relatively inexpensive. Consider purchasing one for your toolbox.

Piston Ring Tools

Special tools are available for fitting and installing piston rings and pistons. Two are optional; one is necessary. Optional tools include a ring filer and a ring expander. As for the necessary one, you can't install pistons in an engine without a piston ring compressor.

Ring Filer—A ring filer is nothing but a fancy tool used to increase ring end gaps when fitting piston rings to their bores. It makes the job of filing ring ends much easier, faster and very neat. But done right, a fine-tooth flat file works just as well. Consider this quality tool from Sealed Power, manufacturer of Speed Pro rings, a "nice-to-have" tool, but not necessary.

Ring Expander—A piston ring expander is used to spread, or expand, a piston ring so it will go over a piston and into its groove. Although it's not necessary that you have one, a ring expander is inexpensive and can save you a lot of money. It prevents your thumbs from being turned raw by the sharp edges of ring ends as you spread the rings. It can also help prevent breaking a ring, so it is bordering on the foolish not to add a ring expander to your tool collection. So forget your thumbs. Weigh the cost of the ring expander to that of a complete ring set. Get a ring expander.

Ring Compressor—A piston ring compressor is a tool you must have to assemble your engine. A ring compressor forces the rings into their grooves so the piston can be slid into its bore. Without one you simply can't install pistons. You must now decide on which type of ring compressor to buy. You have three choices: First is the band type ring compressor with an integral ratchet mounted on the side. It's tightened with an Allen-type wrench that fits in a socket in the

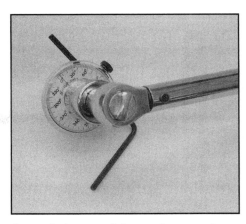

Some bolts used in modern engines, particularly head bolts, are torqued to a specific value, then rotated through a specified angle such as 50°. To do this accurately, you'll need an angle gauge.

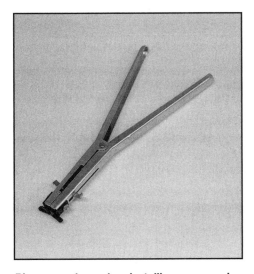

Ring expander makes installing compression rings much easier and greatly reduces risk of breakage. Inexpensive expander adjusts to spread needed to install ring.

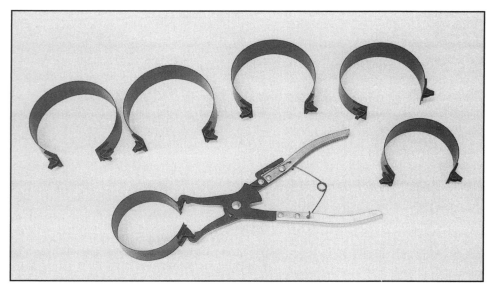

Ring compressor is necessary for installing pistons. Ratchet-type compressor from Sealed Power has interchangeable bands for installing different-diameter pistons.

Loading sleeve is a favorite engine assembly tool in race shops. Rings are compressed as rod-and-piston assembly is pushed into bore. Small end of tapered ID must match bore to ensure rings have smooth transition as they enter bore.

ratchet. Although the least expensive, it works very well. But my favorite ring compressor uses a band that's tightened and held in position with ratcheting pliers that hook into lugs on the side of the band. The last type of ring compressor is a loading sleeve. It has a tapered ID that matches the bore size at its small end. This type of ring compressor was first used in the automobile industry for installing rod-and-piston assemblies on the assembly line. No ratchets or pliers are needed. As the piston is shoved into the tapered sleeve, the rings are gradually compressed as they move toward the smallest section of the taper just as they enter the bore. Very fast. Unlike the other compressors, tapered sleeve-type ring compressors must be used for a single bore size, not a range of sizes. Don't get this one unless you're a professional engine builder.

Number Dies

Also called *steel stamps*, these inexpensive tools are used for permanently numbering metal parts such as connecting rods, their caps and main bearing caps. Numbering such parts is a must before they come apart if they haven't already been numbered. If you elect not to purchase steel stamps, you can improvise by using a *prick punch*—a small center punch—to make a number of punch marks that coincide with the number, or one mark for number 1, two marks for number 2, ... If you elect to purchase steel stamps, get the 1/8-in./3.2mm size. They are big enough to see, but not too big.

INSPECTION TOOLS

To do a successful "stand-alone" engine rebuilding job, you'll need some inspection tools. The good news is you should be able to get along nicely with only a few of these tools providing you can get some help from a friend or your engine machinist. If, on the other hand, you plan on doing all of the inspection work and your goal is to build a blueprinted high-performance engine, you will need an array of tools. Let's take a look at each type and their application so you can make the judgment on whether to purchase or not. Let's start with the most versatile measuring tool, the caliper.

Calipers

Although not the most accurate tool, dial calipers are the most versatile and inexpensive measuring tool you can buy. Although not as accurate as a micrometer, a *dial caliper* can be used for measuring outside dimensions, inside dimensions and how far one parallel surface is from another. Examples of each are crankshaft bearing journal diameter, cylinder bore diameter and valve spring installed height, respectively. Also, a dial caliper can be used to measure a wider range of dimensions than other measuring instruments. A dial caliper with a 0 to 6 inch range is the most common. I recommend you buy this one.

The predecessor to the dial caliper is the *vernier caliper*. As shown in the photo, a dial is used in place of a vernier scale for reading 1/100's and 1/1,000's of an inch. Although a vernier caliper cost less than a dial caliper, I recommend that you pay a little bit more for a dial caliper. It is much easier to read, particularly if lighting is a problem or your sight is not perfect. Put another way, the vernier type caliper is just plain hard to read. Regardless get a vernier caliper if you can't afford a dial caliper. You won't know how you got along without one once you use it.

Micrometers

Commonly called *mikes,* a micrometer is the most commonly used measuring tool in an engine building shop. The best mikes can be read within 0.0001 in., or 1/10,000 of an inch. There are two basic types of micrometers: outside mikes and inside mikes.

Outside Mikes—Outside micrometers are for measuring outside dimensions such as valve stems and crankshaft journals or material thickness such as valve spring shims or head gaskets. They measure in a 1-inch range starting from zero, one, two, three, four . . . and are designated 0-1 in., 1-2 in., 2-3 in., 3-4 in., 4-5 in. . . . , respectively. For example, a 3-4-in. micrometer measures 3 and 4 inches and all points in between to within 0.0001 in. So if what you're measuring is smaller than 3 in. or greater than 4 in., you'll need another micrometer.

If you expect to do a complete and accurate check of your engine, you will need at least three micrometers. The ones you'll need depend on the dimensions of the components you'll be measuring such as valve stems, bearing journals and cylinder bores.

Check the critical dimensions of your engine to determine which micrometers you'll need. Chances are you'll need a 0-1-in., a 2-3-in. and a 3-4-in. micrometer if your engine is typical. Once you decide on the ones you need, shop for quality micrometers first, then look for the lowest price. And quality doesn't mean gizmos such as battery-powered digital readout micrometers that hook up to a video display. What you need are basic durable micrometers that will give you an accuracy within 0.0001 in., or 0.001mm if you choose a metric micrometer.

Features to look for include a vernier scale on the hub, a checking standard, a ratcheting thimble and a lock. The vernier scale allows you to read measurements within 0.0001 in. whereas the resolution scale by itself will only allow you to read within 0.001 in. or 0.01mm. Anything in between will be a guess. The checking standard, or gauge block as some call it, is an accurately machined rod that is used to check and set the accuracy of the micrometer. Note: A 0-1-in./0-25mm micrometer won't need a standard since it should read 0.0000 in./0.000mm when the micrometer is closed. The others won't close up, so they will need standards. All micrometers need a wrench for readjusting their accuracy.

A ratcheting thimble is more of a convenience than a necessity. In fact, most professionals don't use them. If, however, you're not experienced with using a micrometer, a ratcheting thimble can be helpful. It allows you to tighten the micrometer just right when you're taking a reading. When

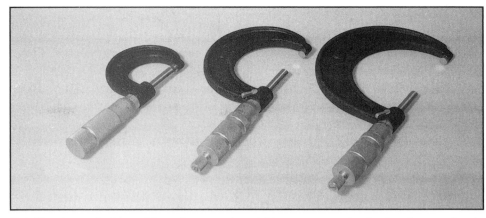

Outside micrometer is used to make accurate outside measurements of valve stems and main and rod bearing journal diameters. Standard is used to check micrometer accuracy. Shown are 0–1-in., 1–2-in. and 2–3-in. micrometers.

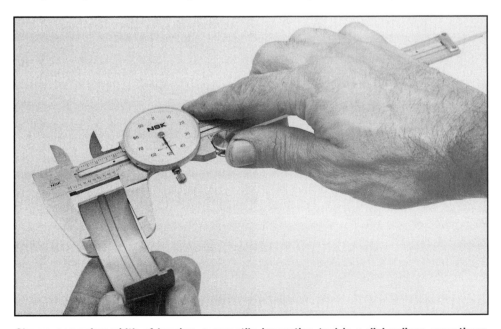

Shown measuring width of bearing, a versatile inspection tool is a dial caliper, sometimes called a vernier caliper. Although not as easy to use, true vernier calipers costs less. Either tool can be used for making outside, inside and depth measurements.

Requiring a special touch, inside micrometers are used for measuring inside diameters such as main bearing and cylinder bores. Extensions are used to change measuring ranges.

the thimble ratchets or slips, the micrometer is tightened correctly. If it is too tight, the micrometer reading will be too low; too loose and it will read high. The lock keeps the thimble from rotating so your reading won't change while you're handling the micrometers.

A final note about outside micrometers: Make sure a ball adapter is included with the 0-1-in./0–25mm micrometer. This attachment allows you to measure curved components such as the thickness of a bearing insert. The ball fits on the end of the spindle and bares against the inside curve of, say, a bearing insert. Just remember to subtract the 0.200 in./5.08mm from each reading or whatever is added by the ball.

For a micrometer that doesn't have a ball adapter, find a ball from an old ball bearing that's about the same diameter as the micrometer spindle. To use, fit the ball to the spindle using a short section of rubber hose that fits tightly over the spindle and ball. Mike the ball and subtract that number from whatever you're measuring.

Inside Mikes—Inside micrometers are used for measuring inside dimensions such as cylinder bores or main bearing bores. But because there are easier ways of measuring inside dimensions with less expensive tools, I suggest you forgo the purchase of an inside micrometer. But because you should be familiar with this tool or you may want one of everything for your toolbox, let's look at the inside mike.

Unlike outside micrometers, you need only one inside micrometer to get a range from one inch to six inches. This range is possible by interchanging various extensions. The problem with inside micrometers is they are extremely difficult to master. You must hold the mike square to the centerline of the bore and directly on center to get an accurate reading. This requires a lot of wiggling and adjusting one way and then another to get an accurate reading . . . then you're not sure. What's frustrating is it seems like every time you take a reading it's different. Consequently, you'll need a lot of practice before you can become proficient with using the inside micrometer.

Telescopic Gauges

Commonly called *snap gauges*, these are transfer gauges rather than measuring instruments. There is no dial or scale from which to read measurements. You'll need outside micrometers to make the final measurement. Even though I have inside mikes, I use telescoping gauges and outside micrometers instead. A typical set includes six telescoping

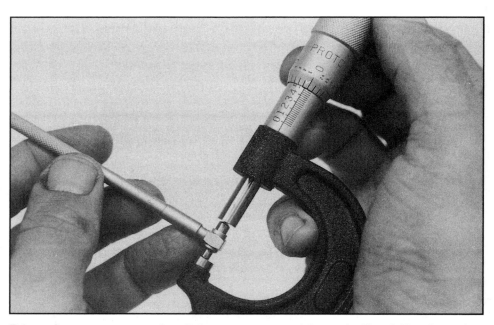

Telescoping gauges, commonly called snap gauges, must be used with outside micrometers for making inside measurements. Although relatively easy to use, you must practice with these transfer gauges in order to get accurate readings.

Dial indicator can be used to check crankshaft end-play, flywheel warpage, crankshaft runout and camshaft lift. Magnetic base works well on flat magnetic surfaces; bolt-on stand is used on irregular surfaces or non-magnetic components such as aluminum cylinder heads. Indicator is being used to find TDC.

gauges. They range from the smallest of 1/2 to 3/4 in.; the largest from 3-1/2 to 6 in.

As shown in the photo, the telescopic gauge is a T-shaped affair. The cross of the T is a spring-loaded plunger with radiused and hardened ends. The handle is turned to lock the plungers in any compressed position. To measure a bore, the first step is to fit the snap gauge in the bore and allow the plungers to spring out against the bore. Next, the handle is lightly tightened while the plungers are against the bore but not quite square to it. I wiggle the gauge a little to center it in the bore before snugging the handle. The final move is to push the handle sideways so it drags the plunger through the true diameter of the bore, compressing it to the true bore diameter. The snugged handle allows the plungers to compress, but prevents them from snapping back out until you can tighten the handle. While the plungers are secured in that position, measure across the plunger ends with the appropriate micrometer to determine bore size.

Small Hole Gauges

To measure a bore that's smaller than 1/2 in. such as a valve guide, a snap gauge won't work. You'll need a small hole, or ball, gauge to fit in the small bore. Available in sets of four, these inexpensive transfer gauges can be used to measure holes or bores as small as 1/8 in. and as large as 1/2 in. I recommend that you add a set of these to your toolbox. You'll need a 0–1-in. micrometer to get readings.

Much easier to use than telescoping gauges, a small hole gauge consists of a split ball with a cone-shaped wedge in the split. After the ball gauge is inserted into a bore to the desired depth, the handle is turned to move the wedge in or out to expand or contract the ball. The gauge should fit in the bore with a slight drag. The gauge is then withdrawn and the width of the ball is measured 90° to the split to determine bore size.

Dial Indicator

You'll need this precision inspection instrument to degree in a cam, check cam lobe lift or blueprint your engine. Dial indicators are also great for checking crankshaft runout, crank end play or piston deck clearance. Although a versatile inspection tool, there are other methods you can use that require the use of less expensive tools if you're only doing a rebuild.

The typical dial indicator consists of a plunger and two dials. As the plunger moves in and out, the dials rotate, indicating plunger travel within an accuracy of 0.001 in. The larger, or outer, dial face is divided into 0.001-in. graduations and the smaller one is divided into revolutions of the larger dial. Each revolution of the larger dial is typically 0.100 in. To provide a convenient starting point, the outer dial face can be rotated on the body of the indicator to zero the pointer. It also has a clamp at the edge to hold

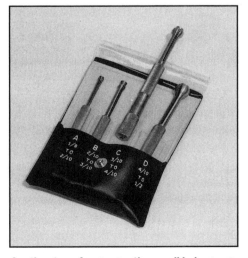

Another transfer gauge, the small-hole gauge is used with a 0–1 inch outside micrometer for measuring small bores such as valve guides. Measurement must be made with micrometer 90° to split in expandable ball end.

Useful for measuring cylinder bores and main bearing bores, a dial bore gauge is the most accurate and easiest to use bore measuring tool. It's also the most expensive. Prior to measuring, gauge is zeroed with an outside micrometer to baseline bore size. Difference is read directly as over or undersize.

the face in place.

Accessories—The most common dial indicators are available in 0.500-in. and 1.000-in. travels. Although more expensive, choose the 1.000-inch-travel indicator because it is more convenient to use. What I consider nice, but not necessary are telltales. These two little pointers on the outer dial OD can be positioned around the dial face to indicate starting or stopping points or ranges for checking tolerances. If the dial pointer does not pass the telltale, you know that what's being checked is within spec.

You'll need two accessories for a dial indicator: a base and a plunger extension. The base is used to support the dial indicator. It can either be bolted or held in place with a permanent magnet. The magnetic base is the most convenient and usually the best to have, but if the part being checked is non-magnetic, such as an aluminum cylinder head or block, it would be better to have the bolt-on base. I say better because you can always bolt a small steel plate to the part being checked. Although not as convenient, you'll then have something to which the magnet will stick.

Between the base and indicator is the adjustable part of the base. As shown, it's either a combination of rods and clamps or a snakey looking affair that clamps solid once the desired position is achieved. I prefer the rod and clamp type base because it's easier to set up and is more solid.

V-Blocks

A pair of cast-iron V-blocks can be considered companions to the dial indicator. They are used to support a crankshaft or camshaft at the ends for checking runout. A crankshaft set in a cylinder block supported by the upper bearing inserts at each end will accomplish the same thing, but you'll need V-blocks for checking a camshaft. Price V-blocks while you're shopping for a dial indicator. Purchase a set if your budget allows.

Dial Bore Gauge

A dial bore gauge is a special-purpose dial indicator for checking bores, and as such they are pretty expensive. The most accurate dial bore gauges have an accuracy of 0.0001 in., or 1/10,000 in.! Less expensive dial bore gauges will check to within 0.0005 in.

As shown, the dial bore gauge consists of a dial indicator at one end of a long stalk and a tripod arrangement positioned 90° to the stalk at the other end. The tripod is made up of a single interchangeable post that goes against one side of the bore being checked and two hardened buttons at the other end that automatically center the plunger in the bore. The gauge plunger is between the two buttons, directly opposite the post. To allow for different bore sizes, the post is interchangeable. This allows you to measure the big end of a connecting rod or the diameter of a cylinder bore.

Reading Technique—To achieve a reading, the gauge is first zeroed. This is done by measuring across the gauge with an outside mike set to the specified bore size and rotating the dial face until **0** aligns with the needle. The gauge is then inserted into the bore to the desired depth and rocked back and forth until the lowest reading is achieved. When the gauge is square to the bore and the indicator needle reverses direction, the lowest reading is read. This may be on the plus or minus side of the zero, indicating an oversize or undersize bore. Because it's so fast and accurate, the dial bore gauge is most helpful while honing bores for fitting pistons.

Although this tool is the easiest and most accurate one to check bores, the cost of a dial bore gauge restricts its use to the professional engine builder or engine machinist. But if you're a very well healed do-it-yourselfer and you feel you must have one, great! Get a dial bore gauge. If you've ever done any engine building without a dial bore gauge, you'll wonder how you ever got along without one. The accuracy is so good you'll drive your

Ribbon gauge is a specialized feeler gauge for measuring piston-to-bore clearances. Gauge and piston are inserted into bore, then ribbon gauge is pulled out with a fish type scale or by someone with a good feel. Pulling tension must be within specified range if clearance is within spec.

engine machinist crazy with the accuracy you'll be able to demand.

Feeler Gauges

Although I have assumed you have a set of feeler gauges, there's one type of which you should be aware. It's the long (12 in.) tempered and polished ribbon gauge for checking piston-to-bore clearance. Thicknesses such as 0.0010, 0.0015 and 0.0020 in. are available. You can use a ribbon gauge as an alternative to the more expensive micrometers, snap gauges, dial bore gauges or whatever for directly measuring piston-to-bore clearance.

Technique—To use a ribbon gauge, install the piston in a clean bore with the gauge alongside the skirt, then pull on the feeler gauge. If there's little to no resistance, clearance is more than gauge thickness. If it is very tight or can't be moved, clearance is less. The gauge will pull out with a "slight" drag if clearance and gauge thickness are the same. An example of a manufacturer's specification is a 6-lb. pull on a 0.0015-in. gauge with a cast piston. Pulling on a fish scale hooked into the hole at the end of the ribbon will eliminate the guess as to the force needed to pull out the gauge.

Plastigage

As with the feeler gauge, this small strip of colored wax is used to measure assembly clearances directly. This is done by compressing the wax strip between the installed bearing and bearing journal, removing the bearing cap and measuring the compressed width of the Plastigage with a scale that's at the edge of the paper sleeve it comes in.

Even though experienced engine builders turn up their noses at it, Plastigage is a handy checking method. Use it to confirm bearing oil clearances. If a clearance doesn't check out right, make sure you have the right bearings and remeasure the bearing journal with a micrometer.

Purchase one sleeve of green Plastigage. It has a 0.001-0.003-in./0.002-0.008mm range. Make sure it is fresh, not hardened by age from sitting on the shelf. Hard Plastigage makes oil clearance readings too low because it doesn't squeeze out as wide as it would otherwise.

Machinist's Blue

Blue dye, which is used by the gallons in machine shops and fabrication shops for laying out parts, is useful for various engine building operations. One of the most common engine building uses for bluing is to apply this quick-drying liquid to valve seats during three-angle jobs, page 91. After top and bottom cuts are made, the valve seat stands out in blue contrast so seat width and diameter are much easier to see for measuring. Similarly, outlining combustion chambers in the process of modifying them, laying out port runners at manifold parting faces or clearancing cam bearings are other examples of application for machinist's blue. As for where to get it, machinist's blue is available at all machinery supply houses and some tool or engine parts stores.

Scribes & Dividers

Both scribes and dividers are

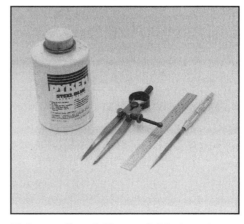

Bluing, scribe, dividers and 6-inch scale are useful for checking valve seat work or making combustion chamber modifications.

Available from camshaft manufacturers and high performance engine parts suppliers, a degree wheel is a must for checking camshaft against the manufacturer's specifications. Directions are printed on wheel.

companions to bluing. Once the blue dye is applied and allowed to dry, patterns are scribed on the blued area. The scribe marks—made with either a scribe or dividers—stand out as bright high-contrast lines against a dark-blue background.

A scribe is basically a pencil with a sharp metal point for scribing straight or curved lines. Dividers is a compass with two sharp metal points rather than one for scribing circles during layout work, finding centers or for transferring measurements. Scribes and dividers are available at virtually all tool supply stores at nominal costs.

Steel Rule

For making non-precision measurements and using as a rigid straightedge for laying out straight lines, get a 6- or 12-inch machinist's scale, or steel rule. This used with machinist's blue, scribe and dividers will allow you to do minor inspection work, blueprinting and performance modifications. The scale should have 1/32-in. and 1/64-in. divisions on one side and 0.100-in. and 0.025-in. divisions on the other. Scales with metric divisions are also available.

Degree Wheel

When doing blueprinting or performance work, you'll need a degree wheel. Use it to ensure the camshaft is ground and installed to the manufacturer's specifications. Degree wheels are offered by most aftermarket high-performance engine parts manufacturers, particularly those who specialize in camshaft and valvetrain components. Choose a degree wheel that is easy to install, has clear directions—preferably on the degree wheel itself—and with numbers that are easy to read. A pointer may or may not be offered, but one cobbled from heavy wire such as welding rod will work fine. When degreeing a cam you'll also need a dial indicator for checking lobe lift, a piston stop or dial indicator for finding Top Dead Center and light-weight checking springs. These items are included in complete cam checking kits offered by some aftermarket manufacturers.

Burette

You'll need one of these scientific measuring instruments if you plan on cc'ing—measuring—the combustion chamber volumes of your engine. Available in different volumes, get a 100cc burette. A 50cc burette won't completely fill the typical combustion chamber, making it inconvenient to use. A stand to support the burette would be helpful, but not necessary. You will, however, need a 1/2-in. thick clear acrylic plate about 8 in. square for covering the combustion chamber.

As for where to get a burette and related accessories, you have several options: If you know a race engine builder, talk to him first. If you don't have this option, visit your local high school or community college. A physics or chemistry instructor may be able to point you in the right direction. Check also high performance parts mail-order catalogs for cc'ing kits.

PREPARATION TOOLS

Files

There always seems to be the need for a file. They are used for enlarging holes, deburring, radiusing and chamfering edges. The problem is one file won't do. You'll need a specific file for a specific job. Three should satisfy most of your filing needs: 10-inch flat, half-round and round, or rat-tail, file, all with smooth or bastard cuts. For jobs where removing metal faster is important, the bastard cut works better, but the finish will be rougher.

Volume of irregular "containers" such as combustion chambers is determined with liquid measured from chemist's burette. Low viscosity liquid such as kerosene is used to fill volume sealed with clear acrylic plate.

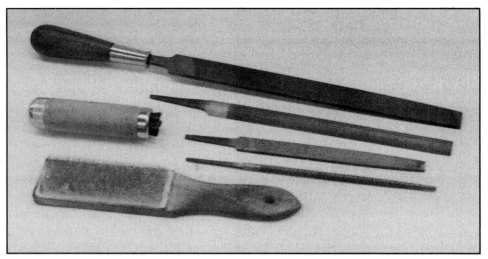

Assortment of files in cut and shape are used to smooth rough surfaces and break sharp edges. Handle installs on tang to make filing easier and safer; file card is for cleaning file teeth.

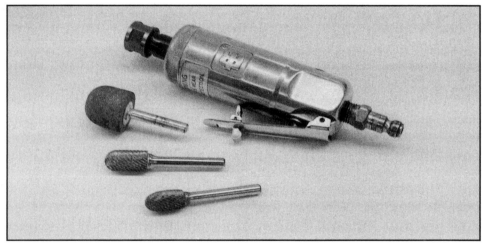

Pneumatic die grinder with assorted carbide burrs, abrasive stones and paper rolls are a must if you intend to make serious cylinder head modifications.

Accessories—Two accessories you need for your files are handles and a file card. For one thing, handles make files much easier to use. A handle will also keep the file tang—pointed end—from piercing the palm of your hand if you run the end of the file into something. A file card is a short-bristle wire brush for removing metal cuttings from the file teeth. A file should be cleaned periodically because one loaded up with metal cuttings will not cut.

For doing precise filing work, consider a needle-file kit. These small, multi-shaped files are great for such things as cleaning up radii on connecting rods at bolt holes or deburring crankshaft oil holes.

Die Grinder—You'll need a die grinder if you'll be doing cylinder head port work or extensive cylinder block deburring. This power tool will give you the capability to remove a lot of metal in a relatively short amount of time. Don't think you can get by with a Dremel type hobby grinder. For major engine work you'll need a high-speed die grinder with a 3/8-in. collet chuck.

Die grinders are either electric or pneumatic powered. The one you get depends on whether you have a high capacity compressed air system—5-hp compressor is more than adequate. If so, get a pneumatic die grinder. It is lighter, more durable and easier to handle than the electric equivalent. Also, get some Marvel Mystery Oil to lubricate your grinder if the compressed air system isn't equipped with an inline oiler.

To do the actual cutting, you'll need some carbide tool steel cutters—*burrs* as they are known in the industry—or abrasive stones. Use burrs for quick material removal, then stones for finishing. The burrs can be used on cast iron, steel and aluminum. Stones are best if they are only used on cast iron or steel; they load up when used on aluminum. You'll need burrs or stones in various 1/2-inch diameter shapes such as round, cylindrical and teardrop, depending on the shape and accessibility of the areas you'll be working. Be forewarned: Carbide burrs are expensive, so you may elect to use only stones.

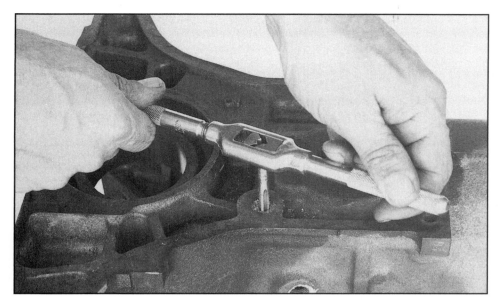
Clean threads with tap to ensure bolts are accurately torqued. Wire brush can be used to clean external threads, but you'll need the correct size and pitch taps for chasing threaded holes.

Taps

Threaded holes must be clean before you can achieve accurate bolt torques. This requires that you chase the threaded holes in the block and heads with hand taps of the same size and pitch. So if you don't have a tap-and-die set and you're unsure about thread sizes, take one bolt of each size to a tool supply store. This will ensure you get taps of the correct sizes. As for the type of tap, there are taper, plug and bottom taps from which to choose. Taper taps are for starting threads in an unthreaded hole, bottom taps are for threading all the way to the bottom of blind holes—holes that don't go all the way through—and plug taps split the difference. Because most threads in an engine are in blind holes, get at least taper taps. These will clean up the threads most of the way to the bottom of a hole. Don't forget a holder, or handle. Although you can use an open-end wrench to turn a tap, a tap holder is much better. ∎

When choosing burr shapes, you must determine the type of cut/s to purchase. Use fine single or double cuts for steel and cast iron. Tip: A coarse single cut burr minimizes the tendency of the cutter to load up with aluminum. To minimize this further, periodically apply wax to the burr when working aluminum. You'll find an ample supply of inexpensive wax in the plumbing section of just about any hardware store. It is used for toilet base sealing rings. Keep a file card or wire brush in reach to remove material from a burr as it loads up.

Grinding stones can be used for all porting and polishing needs. It just takes longer to remove material with them as opposed to burrs. And because they are abrasive, they are also dirtier.

To achieve that highly polished look, get some paper rolls. Not so much for removing material, they put that high polish on a surface that's so desirable for keeping carbon from sticking to a surface such as a combustion chamber. You'll also need a mandrel. It chucks in the grinder; the paper roll is screwed on to the end of the taper threaded mandrel.

Protect Your Eyes—When doing any metal work, particularly high-speed metal removal, always wear eye protection. A die grinder turns up to 25,000 rpm, throwing hot metal and abrasive particles in all directions. Don't risk permanent eye injury. Wear a full face shield or goggles to protect your eyes from injury.

Handy item for organizing parts during teardown or assembly is a silverware tray. Compartments are as useful for separating parts or tools just as they are forks and knives.

ENGINE TEARDOWN 3

Engine rebuilding mistakes begin right here—at teardown. Don't make the first one by getting your engine apart and spread out all over the place as fast as possible without regard to anything else. Tearing an engine down is the first inspection step in engine rebuilding. Consider, too, that if you don't intend to continue with the rebuilding process after teardown, don't start the process yet. An engine is better protected in its assembled, greasy grungy state than with its innards scattered about. I've seen too many engines ruined because of parts being lost or rendered useless by corrosion.

When disassembling your engine, take your time. Those old bearings and gaskets you'll eventually toss out can provide valuable clues as to the condition of engine parts. The same goes for the pistons, bearing caps and bearing journals. You'll quickly be able to discover what was making those internal noises, high oil consumption, loss of power, low oil pressure or low compression—whatever it was that made your engine scream, "I need to be rebuilt!"

First, organize a plan of attack. Separate hardware according to function. Use containers to store parts

A successful teardown is necessary for a successful engine-building project. A clean area, clear bench top and storage containers makes process much easier.

and label them. For example, keep the exhaust and intake manifold bolts or nuts in separate containers rather than mixing them in with other fasteners. Pay special attention to how the parts went together. This is your last chance to see how they all went together. Above all, don't rely on your memory. Make use of marking pens, punches, photographs and sketches before you disassemble them.

EXTERNAL HARDWARE

If you have impact tools you can remove many of the parts with the engine hanging from a chain hoist or cherry picker, but it's not the best method. This is especially true if you don't have impact tools. You'll end up spending more time wrestling with the engine while trying to break loose bolts. Dancing the jig with several

Remove flexplate or flywheel bolts with engine secured so it doesn't roll over. If you don't have an impact wrench, strike wrench with soft mallet to break loose bolts.

hundred pounds dangling from the end of a chain is dangerous, so get your engine firmly supported before you start removing parts.

Engine Stand or Workbench?

The are two common ways of supporting an engine while rebuilding it: with an engine stand or workbench.

An engine stand can support a complete engine during teardown and assembly. The three or four wheels that support the stand allow you to roll the engine around. Additionally, the engine can be rotated 360° around its longitudinal axis and locked in any position. This makes it very easy to work on.

The one problem with an engine stand is cost—they are expensive. Equipment rental outlets will rent them for a weekly or monthly fee. Or if you're lucky, a friend may have one you can borrow. Either way, consider the rental a good investment.

Another way to support your engine is on a sturdy workbench. Many engine builders use benches. These are low in the middle and have drains at their centers. If you opt for a bench, just be sure it will support your engine and the weight of tools.

Clutch

If you are going to use an engine stand, remove the clutch and flywheel or converter flexplate now. In the case of the clutch, match-mark the pressure plate cover and flywheel with a small center punch so you'll have a reference for reinstalling the clutch in the same position.

Don't remove any pressure plate bolt completely. Back one out a full turn, go to the next one and do the same . . . Repeat this process until there is no load on the clutch disc. You can now remove the bolts, but be careful. Be ready to catch up to 50 pounds of pressure plate and disc so they won't fall to the floor where your toes are. And be careful not to get your greasy hands on the disc friction surfaces. If it's reusable, a big greasy fingerprint can cause the clutch to grab and chatter.

If the pressure plate is fitted over dowel pins, it may not come off easily. If this is the case, gradually work the pressure plate off each dowel with a screwdriver.

Drain Liquids

To avoid a big mess, drain the crankcase oil and coolant if you haven't already done so. Teardown is going to be messy enough anyway.

It's simple enough to drain the oil. Just remove the drain plug and let the oil drain into a pan. While you're at it, remove the filter and drain it into the pan, too. This will get most of the oil out of your engine but not all of it. There are plenty of nooks and crannies in an engine to keep all of the oil from draining out, so be forewarned.

As for the coolant, it's a little more difficult. Most of it should've drained during engine removal, however there will be some residual coolant left in the engine. To complete the draining job, either remove the pipe plugs and core plugs—commonly known as freeze plugs or *Welch* plugs—from the sides and maybe both ends of the block. Not all blocks will have pipe plugs.

Flywheel or Flexplate

Once the clutch is off you'll have a clear view of the flywheel. If an automatic transmission was used, a flexplate will be in place of the flywheel.

There's not much difference between removing the two except for the additional weight of the flywheel. To remove either one you'll have to remove the mounting bolts—they'll be tight. If you have an impact wrench, now is the time to use it. If not, get out your 3/4-in. drive breaker bar and the appropriate socket.

To break the bolts loose with a breaker bar, you'll need to keep the crank from turning. Do this with another wrench at the nose end of the crank on the damper bolt. You could also enlist a friend to help. He'll need a pry bar or the biggest screwdriver from your toolbox placed between two starter ring gear teeth. One of the bellhousing dowels or a bolt threaded back into its hole will give you something to lever against as you

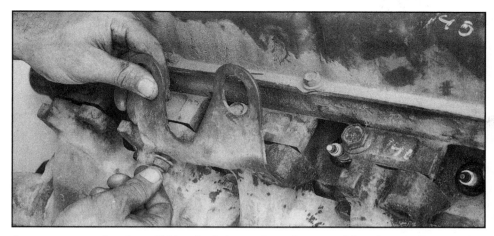

As you remove exhaust manifolds, check for cracks and note position of any baffles or lifting lugs. Such hardware should be replaced, particularly if engine project is part of a vehicle restoration project. Photos from all angles are useful at assembly time.

break loose the flywheel or flexplate bolts with a breaker bar.

Another option is to use a flywheel holding tool such as the one sold by Mr. Gasket. It is nothing more than a short channel with a slotted hole at its center. The tool engages the starter ring gear teeth and is secured to the rear face of the block with a bellhousing bolt.

Engine Plate

An engine plate may be the next item you encounter. This thick piece of sheet metal is sandwiched between the rear face of the engine block and clutch or converter housing. If your engine has one, pull it off of the dowels and set it aside.

If you're going to use an engine stand, now is the time to mount your engine to it. Position the engine on the stand so its center of gravity is close to the stand's pivot axis, or about 4 inches above the crank center.

Exhaust Manifolds & Other Things

If you haven't removed externals such as the exhaust manifolds, starter, fuel pump, motor mounts, carburetor, water pump and distributor, do it now. Set them aside with their mounting fasteners in a clean, dry storage area so you can deal with them later. You may want to restore, rebuild or replace them, but let's concentrate on the job at hand.

GETTING INSIDE

First remove the valve covers—cover if yours is an inline engine. After you've removed the retaining bolts or nuts, pop the covers loose from the cylinder head. It's possible to do this by simply inserting the blade of a screwdriver under one of the corners and striking the handle with the butt of your hand. If it won't come off this way, avoid prying under the valve cover flange in one spot. This will bend it. Instead, work around the flange by prying a little at a time until the valve cover comes loose.

Intake Manifold

Unless you have an engine such as the big-block Ford or 2.8L Chevy V6, you can now remove the intake manifold. On these engines the intake manifold forms the inboard side of the heads. With this design, the pushrods go through the manifold rather than the heads. This means the pushrods must be removed before the intake manifold can come off.

To remove the pushrods, read ahead on how to remove them and return here to continuing your disassembly.

After you've removed the intake-manifold bolts, break loose the manifold. It may seem as though the manifold has grown to the block and heads. Start by prying in several locations between the block and manifold with a large screwdriver. If that doesn't work, stop. Before going any further, check that you've removed all the bolts. Assuming all is OK, lightly drive a small chisel between the block and manifold. Do

Intake manifolds can be ornery. Before you resort to breaking loose manifold as shown, double-check that you've removed all bolts. Be very careful not to damage gasket surfaces, particularly head-to-manifold surfaces.

Some manifolds that seal top of block incorporate manifold-to-head gaskets with a baffle. If your manifold doesn't seal top of block, a separate baffle will be used. Remove it.

Remove individual rocker arms by first removing each hold-down nut or bolt. Regardless of valve-train setup, all parts must be kept in order and together so they can be installed as they came apart at assembly time. Machined parts that wear together must be kept together.

Keep rocker arms and mating pivots together by stringing them on a wire in pairs. Pushrods can be kept in order by pushing them through holes poked into bottom of old shoe box. Label box according to cylinder numbers.

this in several locations at the front and back of the engine. It shouldn't take much before the manifold pops loose. Lift it off and set it aside. If your engine uses a combination baffle/gasket, remove it too. It can be discarded.

Rocker Arms & Pushrods

There are two basic types of rocker arm setups: shaft mounted and individually mounted. There is a distinct difference in how each is removed. But before you can remove a pushrod used with either type of rocker arm setup, the rocker arm must either be removed or shifted out of the way.

Shaft Mounted—Let's tackle the most difficult rocker arm setup first—shaft mounted. To remove the pushrods with this type of setup, the complete rocker arm assembly must come off first. Before you get in a hurry, be prepared to keep all rocker arm components in order and marked so you'll know on which head they were originally installed. This is very important.

Back off each shaft mounting bolt one turn at a time to prevent bending the shaft. Continue this until all rocker arms loosen, then you can remove each bolt. Once they are all out, lift off the rocker arm-and-shaft assembly and set it aside for now. Store bolts with the assembly in their original locations. Remove the other rocker arm-and-shaft assembly in the same manner. You'll inspect and, if required, recondition the rocker arm assemblies later.

Single Mount—For single-mounted rocker arms, back off each mounting bolt or nut so that the rocker arm can be pivoted to the side so it's off the pushrod and valve stem tip. Unlike shaft-mounted rockers, you can leave these in place while you remove the pushrods.

Pushrods—Remove the pushrods. Keep them in order. Take a section of cardboard, and mark it **RIGHT** and **LEFT** for the respective sides and **FRONT** to indicate front of engine. Just punch the lifter end of each pushrod through the cardboard as you remove them from front to rear.

Lifters

If for some reason you intend to reuse the old lifters, conventional flat tappet or roller, you must keep them in order. Also, they must be used with the *same* camshaft in the *same* engine

Mushroom lifters shown on cutaway inline engine must be removed from camshaft side of their bores due to larger diameter foot. This means the camshaft must come out before lifters can be removed.

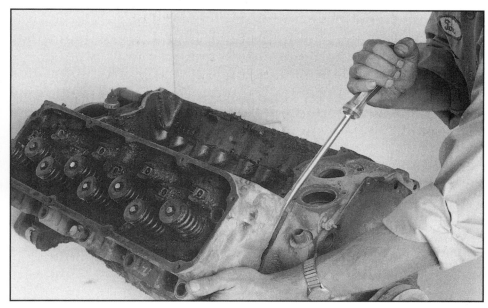

Except for two loosely installed bolts, make sure all head bolts are removed before you try to break head loose from block. Lever off head with breaker bar inserted into intake port; use block of wood with aluminum heads. Take care not to damage gasket surface if you wedge between block and head.

and each lifter must be installed on the *same* lobe. Additionally, never install old lifters on a new camshaft. Conversely, new lifters can be installed on an old camshaft providing it is in good condition.

To store lifters and keep them in order, professional engine builders use lifter trays. These are simply aluminum plates with two rows of eight counterbored holes. You can accomplish the same thing with two egg cartons taped end to end and labeled **FRONT**. You'll just have four pockets on each side you won't use in the case of a V8. For an inline six, you'll just use one row of 12. If your lifters are the roller type, store the guide plates or bars with the lifters.

An engine that's clean inside shouldn't present you with much trouble with removing lifters. Don't use anything that will damage the outer surface of a lifter. This will render it useless. Instead, use a magnet or small screwdriver to catch the retaining ring groove in the lifter ID. Pull the lifter up and out of its bore and store it in order before removing the next one.

If varnish buildup around the bottom end of the lifters makes it too difficult to remove them, spray some carburetor cleaner on each lifter and pull each one out as far as possible. This may expose the relief that's midway down the lifter body and give you something more substantial to lever against. Use a screwdriver to lever with. If this doesn't work you'll have to push the lifters out the bottom of their bores. Before you can do this though you'll have to wait until the bottom end and camshaft are out of the way.

Cylinder Heads

How you remove them depends on whether they are aluminum or cast iron. For aluminum heads, remove the bolts in the reverse order of their torque sequence. Doing otherwise may result in warped cylinder heads. This usually means alternating from one end of the head to the other loosening the bolts as you work toward the center. As for cast-iron heads, the loosening sequence doesn't seem to matter.

Once you have all head bolts out, you can remove the heads. Before you do this, though, make sure the engine-stand neck is secured or the engine is blocked up so it won't roll over due to the off-center weight shift when one head is removed. Also, thread two bolts back in each head two turns or so as a safety measure. This will keep the head from falling off the block and onto the floor where your toes are when it breaks loose.

Cylinder heads can be stubborn to break loose. Once accomplished though, the rest of the job is relatively easy. To break loose a head, first check for an overhanging corner at either end of the head. If you find one, pry under the corner with a pry bar or large screwdriver. If that's not an option, stick a hammer handle or 2x2 block of wood into one of the intake ports and lever off the head.

Once it's broken loose, remove the two "safety" bolts and lift off the head. Using the same method, remove the other head. Mark the heads with a center punch or write on a tag or some tape so you'll know what side it came off of. If shaft-mounted rockers are used, loosely install them on the head for now.

Head Gaskets—Before you

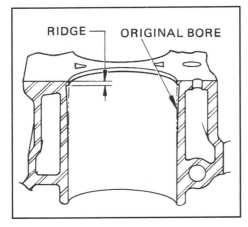

Maximum bore wear occurs where top compression ring stops at TDC. Unworn portion of bore from there to deck surface is the ridge. Remove it before you remove piston.

Removing ridge with ridge reamer. This will allow piston to come out top of bore without ring hanging up and damaging ring land. Removing ridge is a must if you intend to reuse pistons.

proceed further, now is a good time to make your first inspection. This is particularly true if a compression or leak-down test indicated excess cylinder leakage.

Look for discolored areas around each bore. Gaskets are prone to "blow" between bores and from a bore a water passage. If a compression test or leak-down test showed two adjacent cylinders to be down on compression, look for the first. If coolant loss was a problem look for the second.

Remove Ridges

If you're planning on reusing the pistons, the ridges—unworn top of bores—must be removed. If you don't, there's a good chance you will damage the ring lands when the rings catch on the ridges while you push the pistons out the tops of their bores.

If your engine bores aren't worn too badly, you can probably remove the carbon buildup on the ridges by scraping it off with a screwdriver or gasket scraper and pushing out the piston-and-rod assemblies with no damage to the pistons. An easy way to determine this is to run a fingernail up against the bottom of each ridge in several locations around the bore. If your fingernail catches on a ridge, it should be removed.

To remove the ridge, you must have a ridge reamer. If you can't borrow one, rent one from your local tool rental. Run the piston down out of the way and insert the reamer in the top of the bore. Expand the reamer according to the directions that should've accompanied it. Cut the ridge only far enough to match the bore, no more.

Oil Pan & Internals

Take these off before you remove the damper and front cover. Roll over the engine on your bench or engine stand, but be prepared for some coolant to pour out.

With the pan pointing up and all of its bolts out, break the pan loose with a sharp rap or two against a corner with a rubber mallet. If that doesn't work, pry gently under its flange in several locations with a large screwdriver or pry bar. Don't overdo it. Distort the flange and the pan may never seal again.

Now that the oil pan is out of the way you have a view of your engine's bottom end—crankshaft and connecting rods. You'll also see the oil pickup and pump. In rare occasions there may also be a windage tray, or baffle, blocking your view of the rods and crank.

A few factory high-performance engines and some that have been modified are fitted with windage trays. This is nothing more than a formed section of sheet metal that fits tightly to the bottom of the crankshaft. Its purpose is to "scrape" oil off the crank as it rotates and shield it from oil mist and spray, thus minimizing power loss due to oil drag on the crank and rods. Windage trays usually attach to a few of the main bearing cap bolts via studs and nuts or bolts.

First to come off the bottom after the pan is the oil pump and pickup assembly. Unbolt the pump from the bottom of the block. Some pickups are also attached to the block. If that's the case, unbolt it, too. Lift them out as an assembly. Sometimes the oil pump driveshaft will come out with the pump. If it didn't lift it out and store it with the pump and pickup. If your engine has one, remove the windage tray. Store it along with its attaching hardware.

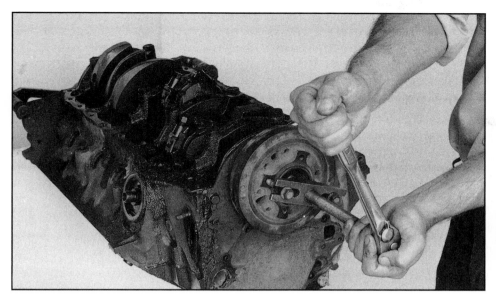

Damper is removed by pulling on bolts threaded into inboard portion of damper. Don't use gear puller to remove damper. It will pull off outer damper ring, rendering damper useless.

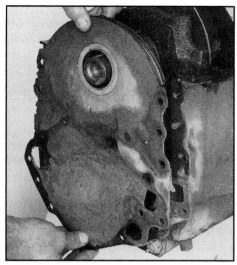

With damper out of the way, front cover can come off. Loosen cover by lightly tapping with a soft mallet. Note dowel that positions sheet-metal cover to front of block.

Water Pump, Damper & Front Cover

Turn your attention to the front of the block and remove the water pump. After you've removed the mounting bolts, break the pump loose from its gasket(s). Do this by bumping the nose of the pump with the butt end of a hammer handle. Once it's off, store the pump with its bolts.

Next on the list is the crankshaft damper. You'll need a steering wheel type puller to do this job, not a pulley puller. And definitely don't attempt to pry or drive off the damper. There's a rubber bond between the outer ring and inner hub of most dampers that will be destroyed if you don't use the right removal technique.

Now is a good time to use an impact wrench. If you don't have one, remove the mounting bolt from the crankshaft nose using a breaker bar and socket. Keep the crankshaft from turning counterclockwise by inserting a wood block between one of the crank's counterweights and the block.

With the damper mounting bolt and washer removed, reinstall the bolt in the nose of the crank. This will give the nose of the puller something to bear against. Fit the puller to the damper with two or three washers and bolts threaded into the damper hub. If you don't have the small mandrels for the center bolt to bear against, find a small nut you don't mind ruining. Smear some grease on the end of the center bolt and install the mandrel or nut between it and the damper bolt head.

Tighten the puller bolt enough to break loose the damper. If it's stubborn, whack the end of the puller bolt with a soft mallet. This should break it loose. Once broken loose, turn the damper bolt and draw the damper off the crank. Once off, remove the bolt and store it and its washer with the damper. Some engines use a spacer behind the damper. Slide it off the crank nose if your engine has one.

With the water pump and crank damper out of the way, you can remove the front cover. It'll either be an aluminum casting or sheet metal, either flat or stamped. Nothing tricky here. Just remove its mounting bolts and break the cover loose from the block. If there's a timing pointer, make sure it and the mounting bolts and washers are stored with the front cover.

Look for an oil slinger that installs behind the damper. It looks like a large-diameter, cone-shaped washer. If your engine uses one, slide the slinger off the crank nose and store it with the front cover. This item must be replaced so that oil that lubricates the timing chain is thrown away from the front crankshaft seal. If it's not reinstalled a large oil leak may occur at the front of your engine.

Timing Chain & Sprockets or Gears

Unless your engine was modified, the timing set will be a chain and two sprockets if it's a V-type engine; if it's an inline engine two gears are frequently used.

In addition to chain versus gear setups, there are various other types of cam related arrangements. For instance, a Buick V6 will have the distributor drive gear positioned ahead of the cam sprocket on the nose of the cam. Most others are behind the number-1 bearing journal built into the cam. Another variation is thrust control. Most camshafts use a thrust

35

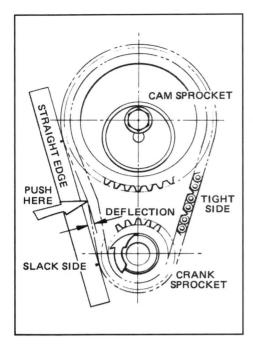

For clockwise-rotating engine, tighten drive side of timing chain by turning crank clockwise; reverse procedure for counterclockwise rotation. Push in on slack side of chain and use straight edge to check deflection. Replace timing set if chain deflects more than 1/2 inch.

plate to handle this. A few use a thrust button at the cam nose that bottoms against the backside of the front cover. Other engines don't use any camshaft thrust control.

Backlash Check—Before you remove the timing set, now is a good time to check it for wear. How you do this depends on whether it's a chain and sprocket setup or two gears are used. Both however involve checking for backlash—simply how much the crankshaft can rotate backwards before the cam begins to rotate.

To check a chain for excessive backlash there's a shortcut. Rotate the crank so one side of the chain is tight; the other side loose and floppy. Lay a straightedge against the loose side, push the chain inward and measure how much it deflects at its midpoint. Consider 1/8 in. OK, 1/4 in. marginal and 1/2 in. junk.

To check a gear drive timing set for backlash, you must measure backlash.

Do this by measuring the clearance between two meshing teeth. You'll need skinny feeler gauges for this job. Or you can measure backlash with a dial indicator. Backlash shouldn't exceed 0.006 in. (0.15mm). You can also use a dial indicator set up so its plunger is tangent with the diameter of the big gear and resting against the face of a tooth. (You'll need an offset on the plunger to do this. With the dial indicator set up, rotate the gear away from the dial indicator and zero the indicator. Now, rotate the gear toward the indicator so all backlash is taken up, but don't rotate the crank gear. Read backlash directly.

Removal—To remove the chain-type timing set, remove the cam sprocket mounting bolt(s), washer(s) and, if your engine has one, the fuel pump eccentric. Once the bolt(s) is out, work the cam sprocket off the cam nose by levering against its backside with two screwdrivers. Before you move it too far, slide the crank sprocket forward a like amount to keep the chain from binding. Work the cam and crank sprockets forward an equal amount—a little at the cam; a little at the crank—until the cam sprocket comes off. Set it and the chain aside and slide the crank sprocket off. If your engine has one, the thrust plate should now be exposed. Remove it and the camshaft will almost be free to come out. In your notes, record anything special about thrust plate installation. Examples are hollow mounting bolt in one location or slots in the thrust plate and, if so, whether they install against the block or sprocket.

To remove a gear timing set, rotate the crankshaft until the access holes in the cam gear align with the thrust plate bolts. Remove these bolts. This will free up the cam so you can pull it out with the gear. Note: Because the gear is usually pressed on the cam,

Chain hadn't jumped timing, but it wasn't far from doing so. Jumped timing frequently occurs when engine kicks back from dieseling or a backfire, advancing valve and ignition timing.

Remove timing set by inching sprockets off together a little at a time. Screwdrivers used as shown usually work. If not, a puller on cam sprocket or crank sprocket may be required to get sprocket to budge.

Before cam can be removed, lifters must be raised in bores so lobes will clear lifters. Do this by rotating camshaft at least one complete revolution. If used, thrust plate must come off to remove cam.

Being careful not to bump bearings with lobes, remove cam from front of block. Sprocket, long bolt or screwdriver at nose of cam will serve as a handle to guide out cam. If crankshaft was removed, position block upside down and support rear of cam with free hand.

you can't remove it while the camshaft is in the engine. This means you'll have to remove the gear with the camshaft.

To remove a pressed-on cam gear you'll have to press it off. You'll need an arbor press or hydraulic press for doing this. Don't attempt to drive off the gear. To remove the crankshaft gear, use a puller and two or three bolts in the threaded holes of the gear.

Camshaft Removal

If you weren't able to remove the lifters earlier, turn the engine upside down or on its side and rotate the crankshaft twice. This rotates the cam once, raising all of the lifters in their bores. To rotate a chain-driven cam, loosely install the old sprocket so you'll have something to grasp for rotating the cam.

Be very careful when removing a camshaft. Banging the lobes of the cam into the cam bearings will damage them. So if you plan on reusing the old cam bearings, be particularly careful.

To start the cam moving forward, using a screwdriver or small pry bar, pry against the backside of a cam lobe while levering against one of the bearing webs. Work the camshaft out of the front of the block. If there's access, reach inside the block and support the opposite end of the cam with your other hand and carefully work it out of the front of the block.

With the cam out of the way, you can remove the lifters. Remember: If you're planning on reusing them, the lifters must be installed on the same lobes. This means you must keep them in order, so store them as described on page 32.

Rods & Pistons

Before you start loosening the rod bolts or nuts, check the sides of the rods and caps at their parting lines for numbers. If you can't find any that correspond to the bores in which the pistons and rods are installed, number them before you loosen the rod caps. Get these caps mixed up and you're guaranteed to have machining expenses that you otherwise may not be faced with.

Numbering—If you have to number your rods and caps, make sure you know how your engine's cylinders are numbered. Inline engines are numbered from front to rear. V-type engines vary. Typically, Ford engines are numbered in sequence starting with **1** at the right front cylinder progressing to the right rear cylinder and continuing from the left front to the left back one. GM, Chrysler and AMC typically number their V-type engines with odd numbers on the left and even numbers on the right. Number 1 starts at the

Badly worn lifter compared to new one; worn lifter foot exhibits a concave wear pattern. Lifter foot should be convex—raised in center.

Lobes and lifters wear together. All it takes is for one lobe of a camshaft to be worn excessively to require replacement. Save cam, though. It may be OK as a core.

Before you loosen the rod caps, check to see if the rods and caps are numbered. Rod numbers must coincide with cylinder numbers.

left front and number 2 starts at the right front. If you are uncertain about your engine's cylinder numbering, look in the factory shop manual. Don't guess.

To number the rods and caps you'll need a small ball-peen hammer and a small set of number dies or, if you don't have access to these, a prick punch. A *prick punch* is nothing more than a small center punch. To mark with a prick punch, make one punch mark for cylinder **1**, two marks for cylinder **2**.... Mark each rod and cap adjacent to the parting line on the thrust side of inline engines and toward the piston's cylinder bank on V-type engines. Typically, the thrust side is on the right side of a conventional rotating engine—on the left side of a marine engine.

To mark the caps, rotate the crank until you have access to the side of a rod. After marking it with the appropriate number or pricks, rotate the crank and do the same to the next rod and cap. Continue this until you've finished marking all the rods and caps.

You're now ready to remove the rod-and-piston assemblies. Rotate the crankshaft so the piston is at BDC for maximum access. You can now remove the rods bolts or nuts. Once they're off, tap the side of the rod and cap with a soft mallet. This will loosen the cap. Pull the cap and its bearing insert off of the bearing journal and set them aside. To protect the bearing journal during rod removal, install sleeves over the rod bolts. Either use the little plastic boots made specifically for this or two short sections of rubber hose. I prefer two six-inch lengths of fuel line hose. Just push them over the rod bolts.

To slide the rod and piston out of the top of the bore, push against the bottom of the piston. A 12-inch or so long wood 2x2 and a hammer work well for this. If there's enough access though, turn your hammer around and, using the head as a handle, push against the bottom of the piston with the end of the handle. As you push against the piston with one hand, have the other one free to catch the piston when it pops out of its bore.

Before you remove the next one, set this rod-and-piston assembly aside and loosely reinstall the cap with the bearing insert halves on the rod. You'll make use of the old bearing inserts later.

Crankshaft

Just as you did with the connecting rods, make sure the main bearing caps are marked as to their position and direction before you remove them. If they aren't marked, mark them. Skip this step and you may not have any way of determining where the caps were originally installed. You'll then have to have your block align bored or honed, page 51.

There may be numbers for position and arrows or triangles for orientation.

Use hammer handle or wood dowel to push against rod or underside of piston with one hand and catch it with free hand. Don't let connecting rod bang against bore.

Once all pistons and rods are out, remove main-bearing caps. Check first that they are numbered. If not, number caps before you remove bolts. Once bolts are out, tap on side with soft mallet to free cap from register or dowels.

Some manufacturers do this for you, others don't.

Numbering Caps—Starting from the front, number the caps **1, 2** and so on with number dies or a center punch in a manner as you did with the rods and caps. Also, make it apparent how each cap is oriented relative to the front of the block. It should be obvious how the rear main is positioned and where it goes, but it may not be with the front cap. As for the main bearing caps in between, they can usually be installed in any position and 180° out.

Numbering to indicate orientation can be done by stamping an arrow on the bottom of the cap that points to the front. Another method is to position the numbers so they would read in normal position with the block standing on its rear face. In other words, place the numbers so their tops go toward the front.

Remove the main bearing bolts. If you have a 406 or 427 Ford, cross bolts are used to reinforce the bottom end. Remove these bolts and their spacers. After removing each of these bolts, store the spacers so you'll know exactly where they were originally installed. They were selectively fitted to the caps and block.

After removing the remainder of the main bearing bolts, remove the caps. This is easier said than done if the caps are in good "shape." They should fit tightly in their registers, making them difficult to pull out by hand. If they are easy to remove, the caps have been distorted from being overloaded such as that caused by detonation.

To remove a tightly fitting cap, insert the threaded end of two cap bolts in it. Using the bolts as handles, lever back and forth on them to work the cap out of its register. If your engine is not deep skirted, you can tap on the side of each cap with a soft mallet while lifting up on the cap. When the cap breaks loose, set it aside with its bearing and move on to the next one. Check to make sure the bearing came out with the cap. Sometimes they'll stay on the crankshaft bearing journal. If it did, remove the bearing half and put it back in its cap.

After you've removed all main caps, carefully lift the crankshaft straight up off its bearings and out of the block. This is sometimes easier said than done. The nose of the crankshaft is easy enough to lift, but the flywheel end doesn't have much to grasp. At this end, you'll have to stick a finger or two in the pilot bearing or counterbore.

A crankshaft stored lying down may warp, so stand it on its flywheel flange. To keep it from getting knocked over, wire the crankshaft to a table leg or nail driven into a wall stud. Organize the bearings according to which main cap and bearing journal they were mated. Join mating halves with some masking tape and record on the tape which journal they were installed on. These and the rod bearings can be valuable during the inspection and reconditioning stages.

BEARING INSPECTION

Before you discard the connecting rod or main bearings take a good look at them; they'll tell you a story. Bearing inserts are made from plated copper lead alloys or lead-based babbitt, both on a steel shell.

If your engine had a lot of accessories mounted on it such as power steering and air conditioning, the front top main bearing insert is likely to be worn more than the others. This is caused by vertical load placed on the crankshaft by the drive belt(s).

This wear is normal because of the way the crank is loaded and should not be a cause for any concern. For the same reason, wear may also show up on the bottom of the center bearings, particularly at the second and third journals. If the bearings are copper lead type, a copper color will show evenly through the tin plating. This makes it easy to distinguish wear because of the contrasting colors of the tin and copper. As for the lead-based bearings, it's more difficult to distinguish wear because of the similar colors of the different overlays.

You should be concerned about

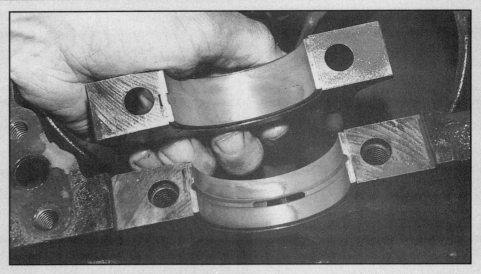

Main caps and bearings tell a story. If caps were loose in registers, chances are engine experienced severe detonation. Wear on grooved upper bearing half is more than on lower half of front insert because of upward force imposed by accessory-drive belts.

uneven wear from front to back on the total circumference of a bearing insert (top and bottom), scratches in the bearing surface and a wiped bearing surface. The first condition indicates the bearing journal is tapered, meaning its diameter is not constant from end to end, causing uneven loading and the resulting uneven wear. Scratches in the bearing surface mean foreign material in the oil passed between the bearing and crankshaft journal. The usual cause of this is dirty oil or debris from failed parts such as metal chips from a wiped camshaft lobe. Serious wear results when a clogged oil filter is bypassed, allowing unfiltered dirty oil to pass through the engine. A wiped bearing surface is usually cause by oil starvation. When lubrication is inadequate, metal-to-metal contact results.

The cause of all these problems must be checked and remedied. Also, any problem you may find with the bearings means there may also be damage to the crankshaft. Consequently you should pay particular attention to the crank journals that had bearing damage to see if the journal was also damaged.

Plugs, Cam Bearings & Things

Your engine should be free of all moving parts. What's left are the oil gallery plugs, cam bearings and rear plug, core plugs and possibly the oil filter adapter. Remove these parts to complete your teardown.

Camshaft Bearings—These are precision inserts that should assume their correct size after they are driven into their bores. Also, bearing bores in the block are machined on a common center line, meaning the bearing bores

Drive out cam bearings with a mandrel, drive bar and big hammer. Mandrel must fit bearing ID, shoulder against bearing shell and clear bearing bore as bearing insert is driven out. Some engines use the same-size bearing inserts from front to back and some don't.

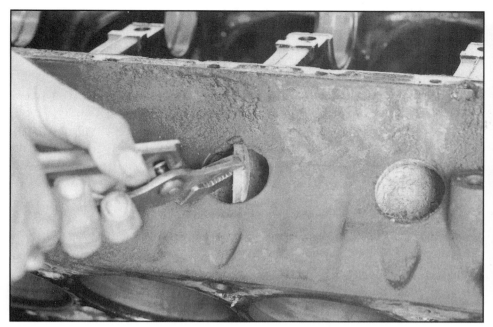

Use large dowel or punch to drive in core plug, grasp flange with pliers and lever plug out of core hole. Condition of core plugs indicate how well cooling system was maintained.

Oil gallery plugs must be removed to ensure that oil passages are cleaned. Not always, but typically pipe plugs are used at back of block and cup-type plugs at front. Remove pipe plugs first. Heat plugs cherry red before you try removing them.

should also have a common center after they are installed. Unfortunately this isn't always the case.

If newly installed cam bearings are not the correct size or are misaligned the bearings must be trued or clearanced by align boring or the bluing-and-scraping method. Either method requires the skill and equipment of an experienced engine machinist. So consider this while you're deciding whether or not to remove the cam bearings.

I am devoting more space than usual to the subject of cam bearings so you'll understand the importance of avoiding damage during teardown and cleaning if you decide to leave them in place for reuse. Be aware that a caustic hot tank solution, so ideal for cleaning steel and cast iron, dissolves soft metals such as that used for bearing material. The good news is the bearings can be saved and the block cleaned if a spray jet tank is used, if the cleaning solution is non-caustic.

Remember, camshaft bearings wear so little that a normal set could outlast two or three engine rebuilds. On the other hand, if your engine had a failure that resulted in large amounts of metal chips being circulated through the lubrication system, cam bearing damage would have resulted—replace them. So, with these points in mind, you'll have to decide now whether you're going to attempt to save the bearings or not.

If you are thinking about removing the camshaft bearings yourself, don't! Unless you have an engine machine shop or access to cam bearing installation and removal equipment and the know how to use it, farm out this job.

Cam bearing removal tools vary in sophistication from a various size solid mandrels that fit snugly the bearing ID to a collet type that expands to fit the bearing. Both types have a shoulder that bears against the edge of the bearing shell but will clear the bore in which the bearing is installed as it's removed. The bearings are removed with a drive bar which centers in the mandrel, or by a long threaded rod which pulls the mandrel. Either type of cam bearing installation/removal tool works fine.

The important thing to remember is to make sure the bearings are installed square in their bores and without damage. I'll take up bearing installation when it's time to do so.

If you have elected to remove your engine's cam bearings, do it now. Remove the cam plug at the rear of the block by driving it out, then remove the bearings. Drive the plug out using a long bar inserted through the bearings from the front of the block. Use the bearing mandrel(s) and drive or pull out the bearings.

Line up the bearings in order of how they came out. Inspect them closely for features such as more than one oil hole, size differences and grooving differences. Note these special features in your notebook.

Core Plugs—These are usually cup-type plugs; round with a wide flange at the periphery. Remove these plugs by driving them into the water jacket and levering them out. To drive them in, use a large diameter punch and a hammer. Pry them out by grasping the plug flange with a pair of

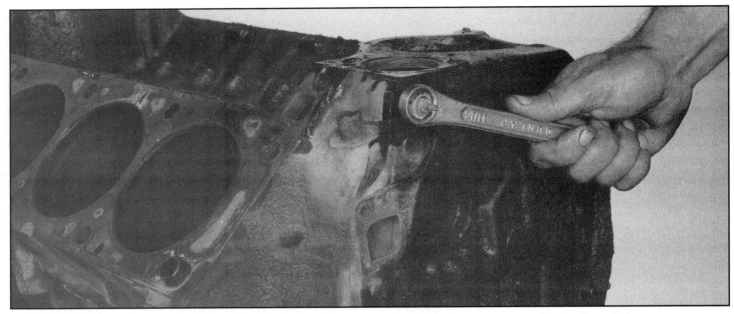

Remove all sending units or other items that may be damaged or seal off areas that should be exposed to the cleaning solution.

water pump pliers. If the plugs were badly rusted, you'll know there was insufficient antifreeze or rust preventative in the coolant.

While you're at it, if core plugs are installed in the cylinder head(s), remove them also.

If pipe type drain plugs are installed in the water jacket, remove them.

Oil Gallery Plugs—To adequately clean your block these plugs must come out. There are two basic types of oil gallery plugs: threaded and cup type. Your engine probably uses both. Threaded pipe plugs are typically used at the rear of a block because of the consequences of an oil leak and, thus, their superior sealing.

As for the threaded plug, there are two types; one has a recessed square or hex head and the other a male square drive. Something these plugs seem to have in common is their difficulty in being removed. If you can't seem to make any progress by using a hex or square Allen wrench or socket extension of the proper size, heat the plugs cherry red—no more—and give it another try. If this doesn't work, don't risk damaging your engine. Have your engine machine shop remove the oil gallery plugs for you. Remember, these plugs must come out in order for you to do a thorough cleaning job.

You can either remove cup-type oil-gallery plugs by driving them out or pulling them out. To drive out oil gallery plugs—the easiest way to remove them—you'll need a steel rod that's a few inches longer than the block and is small enough to fit into the oil gallery. Run the rod through the oil gallery and simply drive out the plug. To pull them out, drill a small hole (about 1/8 inch) in the center of the plug and thread a sheet metal screw in the hole. Clamp on the screw head with Vise Grip style pliers and use a small pry bar or large screwdriver against the plier jaws to lever out the plug. ■

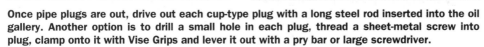

Once pipe plugs are out, drive out each cup-type plug with a long steel rod inserted into the oil gallery. Another option is to drill a small hole in each plug, thread a sheet-metal screw into plug, clamp onto it with Vise Grips and lever it out with a pry bar or large screwdriver.

SHORT BLOCK 4

Major engine building problems usually occur or are caused during the inspection, reconditioning and assembly stages. So if you're like most people you have yet to encounter any problems. Unfortunately problems don't surface until all of your budgeted money has been spent and the engine has been reinstalled and is running. . . or won't run at all.

Engine building goofs are usually the result of being impatient, not using care nor taking advantage of readily available information. Because of this I've attempted to include information you'll need to inspect and recondition your pushrod engine regardless of brand or configuration. Regardless, you should have detailed information on your engine such as that found in manuals and spec sheets from the engine manufacturer or aftermarket parts suppliers. So, with the information in this book and specific information used with an abundance of common sense, patience and a reasonable amount of care, your engine should perform well from the moment it is started.

To blueprint your engine—restore it to specifications within very close tolerances—you must have the manufacturer's, original engine

Block is honed with torque plates and main bearing caps in position to ensure true bores are achieved. Note fluid used to cool and wash bores.

builder's or aftermarket parts manufacturer's specs. You must then inspect each component very closely to determine if any or how much machining or parts replacement must be done.

Note: Before you make a final determination on replacing engine parts or machining them, do a complete inspection job. The reason is simple: The replacement of a part or the machining of it may affect another component or area of the same part. With that said, let's get on with the job.

BLOCK CLEANING

One of the most important engine building jobs you do is clean each component. After you've finished with cleaning each part, remember that it becomes equally important that you keep the parts clean.

Cast-iron block is submerged in hot caustic solution to clean exterior surfaces, oil passages and other exposed areas. Only steel and cast-iron parts can be hot-tanked. Parts should be rinsed with clear water and rust proofed with a water-dispersant oil immediately after hot-tanking. Alternate methods are used for cleaning aluminum parts.

Head Gasket Check

Before you destroy the evidence, inspect the head gasket(s) for signs of combustion and coolant leaks. A coolant leak shows up as rust streaks on the block and/or head surfaces even though one may be aluminum. Black or gray streaks radiating from a combustion chamber or cylinder indicate a combustion leak.

If you find signs of a coolant leak connecting a cylinder with a coolant passage, your engine was probably losing coolant. This is because combustion chamber pressures over pressurized the cooling system and forced out coolant. On the other hand, a compression leak vented to the atmosphere or another cylinder will have shown up only if you performed a compression or leak-down test prior to tearing down your engine. Regardless of whether you found a leak, don't fail to check the block and head surfaces for warpage or other imperfections around the suspected area.

One type of leak to be particularly watchful for is a combustion leak connecting two cylinders. The block or a cylinder head may be notched—cut by the burning air/fuel mixture. Notching occurs when hot combustion gases remove metal as they flow back and forth between cylinders. The effect is similar to the action of an oxyacetylene torch.

Aluminum cylinder heads are particularly susceptible to notching. Although more prevalent with racing engines or those used in boats, notching can also occur with street engines such as those used in tow vehicles. If a cylinder suffered a combustion leak, check for notching after cleaning the head(s). Notch damage can be repaired by welding and resurfacing the damaged area. Although not impossible, welding is much more difficult and expensive with cast-iron heads and blocks.

Commercial Cleaning Methods

How you clean your block can be done using one or more of several methods. Whatever you use, you should include the traditional steam-cleaning method. If you don't own one or have access to a steam cleaner, take a spray can of degreaser and your block, head(s) and crankshaft to the local car wash. Soak the parts with degreaser according to the instructions, then spray off the grime, grease and crud.

The old tried-and-true method, of hot tanking with caustic soda remains very popular with engine rebuilders. Other methods have come along to do the job, too. These include jet spraying and dry cleaning.

Jet spraying or hot spraying is a

Special attention is paid to oil passages. Steam is used also to clean bolt holes and water jackets.

process where the part to be cleaned is placed on a table that rotates. As the block or whatever is rotated, movable nozzles spray a hot solution under high pressure at the part, resulting in a rapid and thorough cleaning job.

Dry cleaning is slower. It's done by placing the part(s) in a cleaning oven or furnace where it is heated for up to six hours. This bakes the contaminants, virtually turning them to dust. It can then be shaken or tumbled out of the block or head and discarded.

Except for steam cleaning, the above methods are commonly used commercially. Whatever method you use though, have as many of your engine parts as possible cleaned at the same time. This includes the block, head(s), front cover, intake manifold, oil pan and rocker cover(s). Whatever method you use, concentrate on the block interior, particularly the oil galleries and bolt holes.

Oil Galleries

To clean the oil galleries, hook a rag on a wire and drag it though the oil galleries. Many supermarkets sell nylon coffee-pot brushes that are well suited to this job. Copper-bristle gun bore brushes work well, too. Team this with carburetor cleaner and you'll be able to do an excellent cleaning job on the oiling system.

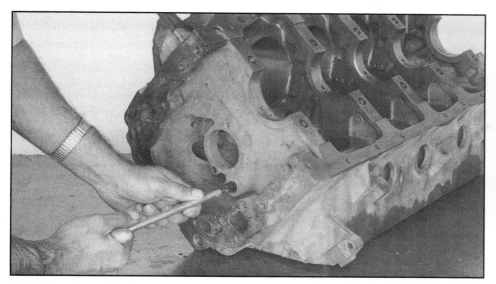

Following steam cleaning run a gun-bore brush through oil galleries to loosen stubborn deposits. Brush should fit tightly in passage. If brush handle is not long enough to reach to opposite end of passage, do one end, then run brush in from opposite end of block.

Putty knife will work, but official gasket scraper is best for removing gaskets from cast-iron. Use a chemical remover, lacquer thinner or alcohol with a plastic scraper to avoid damaging aluminum gasket surfaces.

Scrape Gaskets

Get the worst job out of the way by scraping all of the gasket surfaces. Use a gasket scraper, a tool specifically designed for doing this. Try the job with a putty knife or a screwdriver, then switch to a gasket scraper to find out how much easier it is. Get some spray-on gasket remover, too. You'll save skinned knuckles, a lost temper and loads of time.

Block surfaces that need to be scraped to bare metal are the deck(s), front cover, oil pan, intake manifold and, possibly, a side or lifter-valley cover. Don't stop with the block. Scrape the head and exhaust-manifold gasket surfaces, too.

Chase Threads

After you get the block as clean as you think you can get it, chase the threaded holes. Run a tap the same size and pitch as the original bolt into the threaded hole. This will clean it out all the way to the bottom thread. Threading the tap into the hole should offer only slightly more resistance than running in the correct bolt. You'll be shocked at the amount of debris you'll extract from the threads, even after you've done a "meticulous" job of cleaning.

Tap Types—Do the thread chasing with a bottoming tap, not a taper tap. The bottoming tap cuts a full thread on the starting end rather than being tapered as is necessary for starting a new thread.

The thread-chasing procedure is particularly important for bolts that must be torqued accurately during assembly. The major ones are those for the main bearing caps and cylinder head bolts.

Clean Coolant Passages

After you've cleaned head and main bearing bolthole threads, turn your attention to the water jackets. Remove any loose rust, deposits or core sand. Pay particular attention to the passages connecting the cylinder head to the block. This will ensure good heat transfer and coolant flow between the head(s) and block.

A round file works well for this job, but take care that you don't damage the deck surface(s). A gouge in the

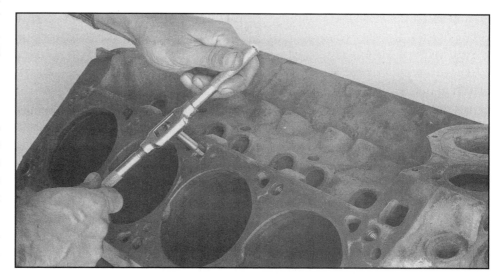

Using correct tap, chase each threaded hole in block. Clean threads will help ensure bolt are torqued correctly during assembly. Remove debris and liquid from bolt holes with compressed air.

Allen Johnson uses orbital sander to clean deck surface of race-engine block. The object is not to remove metal, but to provide clean surface for new gasket. Note needle-type cam bearing in this NASCAR National engine.

wrong place can lead to a head-gasket leak. Give the same treatment to the cylinder head(s).

Compressed Air

An indispensable aid for forcing dirt out of hard-to-reach areas and for drying the block is compressed air. If you don't have an air compressor, a portable air tank with a blow gun will work nearly as well. You'll just have to pump it up periodically.

Prevent Rust

Controlling moisture becomes an increasing problem as you clean more of the parts. Bearing bore surfaces, cylinder bores, valve seats and other machined surfaces will rust from moisture in the air after old oil, grease and sludge are removed.

To protect against rust, coat machined surfaces with a water-dispersant oil immediately after cleaning or machining. Examples of such oil is WD-40 or CRC. Don't use motor oil. Rather than dispersing moisture, motor oil traps it underneath where it can corrode the clean surface.

BLOCK INSPECTION & RECONDITIONING

The first reconditioning step is to inspect the block to determine what must be done to restore it to its original or better condition.

To do a proper inspection job you'll need 3–4 or 4–5 inch inside and outside micrometers, an accurate straightedge and a set of feeler gauges. A set of telescoping gauges will eliminate the need for inside mikes. You may not need the straightedge if the head gasket checked out OK. If the original gasket didn't leak, the new one shouldn't either providing it's installed correctly. Regardless, I recommend that this check be made just to be on the safe side even though you may have to tote your block to the machine shop. This is particularly true if you are blueprinting your engine for maximum durability or performance.

Check Bore Wear

Cylinder bore wear dictates whether a block needs boring or just honing. This, in turn, also dictates whether you'll have to install new pistons . . . no small investment.

There are three common ways of checking bore wear. The best method is to use a dial bore gauge, but chances are you don't have one of these. Next is the inside micrometer or telescoping gauge and outside mike. The next to least accurate method involves using a piston ring and feeler gauges. End gap changes with bore wear—more wear, more gap. A quickie way of checking bore wear involves the use of running a fingernail up under the ridge in several locations! Many rebuilders use this method to make initial estimates of bore wear.

Accuracy of the above measuring methods range from very precise to close enough. Don't consider the fingernail method a measuring technique, though.

Bore Taper—Bores don't wear the same from top to bottom. They wear considerably more at the extreme top portion of the top ring travel and from little to none as the piston goes down the bore. This varying wear creates the so-called *taper*.

High pressure from combustion is exerted against the bore by the top compression ring during the power stroke. This pressure decreases rapidly during the burning of the air/fuel mixture as the piston travels down the bore from TDC, thus decreasing ring-to-bore pressure and the resulting wear. In addition, the bottom of a bore, which mainly stabilizes the piston, is much better lubricated and is lightly loaded by the piston. This is evidenced by the shiny (glazed) upper part of a bore; the lower surface of a bore usually retains its original crosshatch.

Measuring Taper—Measure bore taper by comparing the distance across the bore (diameter) at the bottom versus that at the top, directly below the ridge.

Bores not only wear differently from top to bottom, they wear unevenly around their circumference. Measure across each bore inline with the crankshaft, then 90° to the crankshaft center line. Take several measurements in between and use the highest figure as maximum bore wear.

You won't be able to detect this difference in wear when measuring with ring and feeler gauges; you will with a dial bore gauge, inside mike or an outside micrometer and a telescoping gauge. Instead, you will be measuring average wear. Don't be concerned. Your measurements will be close enough to determine if the block needs to be bored.

Sleeving—One exception to this rule is when one cylinder is damaged or worn past the limit, but the others are OK. In this case, it may be less expensive to have that cylinder sleeved rather than junking a block. It's also possible that your rare or expensive block cannot be bored because it's already to maximum oversize or you must retain a certain bore size. This is a situation where it may be desirable to sleeve all cylinders.

Due to uneven wear around a bore, measuring may not tell you how much a cylinder must be bored to clean it up—expose new metal the full length of the bore. This is because uneven wear shifts a bore's center line in the direction of maximum wear. Consequently, the bore must be restored to its original center line.

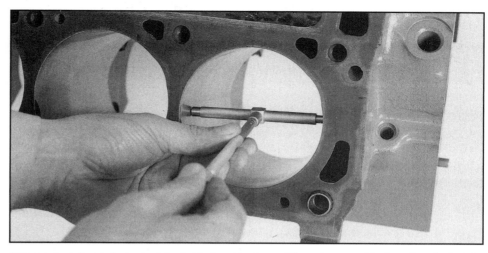

Telescoping gauge is locked to hold plungers in position at a specific location, locked in position, withdrawn from bore, then measured with outside micrometers to determine bore size. Subtract unworn bore size from that having maximum wear to determine taper.

Here's an extreme example: If a bore is worn 0.007 in., and 0.005 in. of this wear is on one side of the cylinder, it must be bored 0.010-in. oversize, or the first available oversize. It's foolish to bore a block more than necessary.

Dial Bore or Micrometer-Telescoping Gauge—When using a dial bore gauge or telescoping gauge and/or micrometer to measure taper, measure the point of maximum wear immediately below the ridge. Because wear will be irregular, take several measurements around the bore to find maximum wear. Determine taper by subtracting the measurement at the bottom of the bore from that at the top. This is automatically done if you're using a dial bore gauge—it is zeroed at the bottom of the bore.

Ring & Feeler Gauge—Although an indirect measurement and therefore less accurate, you can measure bore wear with a piston ring and feeler gauge. Compare the difference between the circumferences of the worn and unworn sections of the bore.

Accuracy decreases the more irregularly a bore is worn. However, it is accurate enough to determine if you'll need to bore and install oversize pistons, or if you can get by with cleaning up the original pistons,

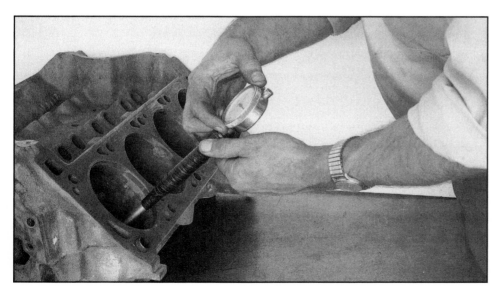

Dial bore gauge is used to measure bore wear directly. Typically maximum wear is found immediately below ridge in line with main-bearing bores in end cylinders and 90° to mains for all others. Check several other positions to make sure, though.

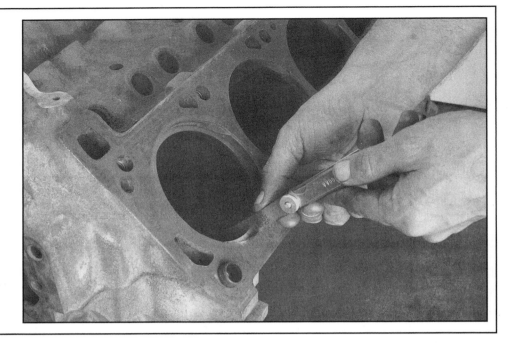

G_2-G_1-ΔG	TAPER
0.000	0.0000
0.001	0.0003
0.005	0.0016
0.010	0.0032
0.015	0.0048
0.020	0.0064
0.025	0.0080
0.030	0.0095
0.035	0.0111
0.040	0.0127
0.045	0.0143
0.050	0.0159

Approximate Taper = 0.30 x ΔG

The difference in ring end gap from the bottom of the bore to that immediately below the ridge multiplied by 0.30 will give approximate taper. Taper can be read directly from chart. In the chart, ΔG is the difference between G2 and G1.

honing the bore and installing new rings.

To use the ring-and-feeler method, place a ring in the cylinder and compare ring end gaps. Measure the gap with the ring at the bottom of the bore, then slide the ring up against the bottom of the ridge and remeasure the gap. Use the same ring, used or new, and make sure it is square in the bore to get an accurate reading.

To square the ring, push it down the bore with a bare piston—no rings. Measure the gap with feeler gauges and record the results. After determining end gap differences, use the nearby chart to determine taper. Taper is approximately 0.3 times ring end gap difference.

How Much Taper Is Too Much?—To decide how much taper your engine can tolerate, ask yourself how many good miles, runs or laps you expect from it. Do you want it to go another 10,000, 50,000 or 100,000 miles . . . all season . . . 500 laps? The bottom line is if you want to build your engine properly, the bores must be straight—no taper. Anything less is a compromise.

If the bores are excessively tapered, new rings won't fix the problem. The rings, especially the top ones, will quickly fatigue and lose their *tension*—resiliency that provides the force they exert against the cylinder walls. Additionally the constant expansion and contraction of the rings as they travel up and down the tapered bore will also accelerate piston ring land wear, further reducing bore sealing.

The story doesn't get any better. Because ring end gap must be correct at the bottom of a tapered bore, the gap is greatest at the top of the bore where compression and combustion pressures are highest. Consequently, there will be further leakage through the wider-than-desired gap. There's more: End gap increases with taper as the miles pile up. So building an engine with any taper, unless it's a "thrasher" is foolish. In this case you can get away with as much as 0.010-in./0.254mm taper. If you are using the ring-and-feeler gauge method of measuring taper, 0.006 in./0.150mm is the number.

If your objective is to end up with a decent rebuild—no more—don't allow taper to exceed 0.004 in./0.100mm. Remember, bore taper gets worse with use, never better. It's still best to start your newly built engine with straight bores and new pistons for maximum durability and performance.

Piston-to-Bore Clearance

All of this talk about whether or not to bore may turn out to be purely academic. The main reason for not boring is usually to avoid the cost of new pistons. This is a valid reason because the cost of new pistons with rings comes close to the cost of sleeving a cylinder. Another situation is where you wish to maintain a certain bore size because it's necessary or desirable for reasons mentioned earlier.

If the pistons are damaged or worn to the point of being unusable, you must replace them whether the block needs bored or not. So, to make the

Glaze breaking is simply restoring cross-hatch pattern to bore surface. Flex-type hone follows existing bore without removing significant amounts of material.

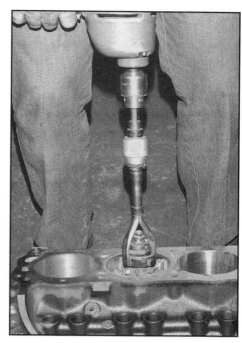

Doing it the old-fashioned way, honing by hand can yield very good results when done by an experienced engine machinist.

final determination as to whether you should rebore, check piston-to-bore clearance. Refer to page 62 for other checks to make before you give the pistons the final OK.

Two methods can be used for checking piston-to-bore clearance. The first is mathematical. Measure across the cylinder bores 90° to the crankshaft center line and about 2 in./50mm below the block's deck surface. Now measure the piston that goes with that bore. Both should be at room temperature—about 70F. If the temperature of these components differs by much, your measurements will not be accurate. Take this into account when holding a piston in your hand. Body heat will warm the piston and cause the aluminum to expand.

Generally speaking, piston-to-bore clearance varies according to piston material and engine application: street, street/strip, racing . . . For instance, forged aluminum pistons require more clearance than cast ones. As for application, a piston for street use only uses less clearance than a street/strip piston regardless of material. A race-only piston requires even more clearance. For this reason you should use the engine manufacturer's specifications when you're using the original pistons. For replacement pistons, use the piston manufacturer's specs.

Piston Knurling—This is a low-budget engine rebuilding method. I don't like piston knurling. It's sort of the engine rebuilding equivalent of repairing a rusted-out body with newspaper and plastic filler. Knurling will work for a short while, but it won't be long that you'll be worse off than before. It just makes a "tired" engine a little less "tired."

Knurling a piston involves rolling a waffle-looking pattern in both skirts. This raises a section of material locally, thus reducing piston-to-bore clearance. The main reason for knurling is to avoid the cost of replacing pistons and boring the block. Don't do it! Replace the pistons and bore the block if piston-to-bore clearance is excessive.

Glaze Breaking

If bore taper is within limits, the next step is breaking the blaze. Check the crosshatch pattern at the base of the bores. This is what you'll restore bore finish to. The purpose of glaze breaking is to restore a honed finish for positive ring break-in and sealing. How coarse the bore finish depends on which type of ring you're going to use.

A glaze breaking hone simply resurfaces the existing bore—it won't remove taper or change the shape of a bore in any way. On the other hand, a precision hone will try to make a cylinder round and straight. It also increases bore size and piston-to-bore clearance—not desirable for glaze breaking.

If you are going to use plain or chrome rings, this operation is necessary. It is optional with moly rings—but desirable because of the ring break-in aspect. Again, glaze breaking is not done to remove material, so a precision hone should not be used. Instead, use a spring-loaded hone or a brush-type hone that follows the existing bore. Before you send your block off to be bored or honed, first check for any irregularities on the deck surface and in the main bearing bores.

Choosing Piston Ring Type

When rings are discussed, it usually concerns the top ring only. And the materials typically discussed are cast iron, moly, chrome or ceramic. But all top rings are cast iron, either plain or ductile. The material that's referred to is the ring facing, or the portion of the ring that runs against the cylinder wall. As the saying goes, "It's where the rubber meets the road." There are other factors to consider that you don't hear much about such as ring width,

Torquing main bearing cap bolts in place is critical before honing. Force from bolts actually distorts bores ever so slightly, but enough to show up as uneven bore wear if caps are not in place during honing.

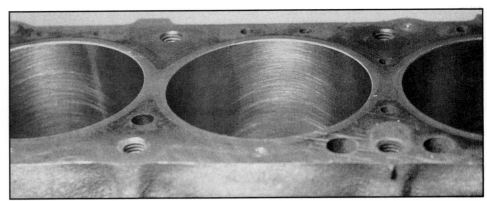

Cross-hatch pattern is particularly important if chrome-faced rings are to be used. Stroke speed and hone rpm determines crosshatch angle, which should be between 20° and 60°.

Deck plates are installed to simulate bolted-on heads. As with main caps, head bolts also distort bores. Unlike cylinder heads, deck plates allow access to cylinder bores for honing. Cross-hatch is restored to bores in freshening-up process between races. Minimum amount of material is removed from bores.

twist, tension, facing shape, cross section and end gap treatment. But these are of little concern unless you're building a specialty engine. Otherwise confine your ring decision process to facing material only.

If the cylinders will be honed but not bored, use cast iron rings. The bores must be "perfectly" straight. More than 0.001-in./0.03mm taper will prevent other ring types from seating, resulting in excess blowby and oil consumption. Cast iron piston rings are very forgiving when used in re-ring jobs where the cylinders were only honed. On the other hand, if you rebore your engine, the choice of ring material is flexible. You must still consider variables such as initial cost, durability, application, fuel used and conditions under which the engine will be operated. Let's consider the first variable—cost.

If cost is a prime concern, use plain cast iron rings, regardless of bore condition. But if the engine is for heavy-duty use such as trailer towing or other conditions where combustion chamber temperatures will be high, at least go to a premium ductile cast iron in the top groove. It is stronger, more fatigue resistant and can operate at higher temperatures.

For nearly all other applications, go with moly or ceramic ductile cast-iron rings if the bores are straight. Not only is the facing tough, the plasma coated, or filled, moly or ceramic facing is very porous, enabling the ring to carry most of its lubricating oil. This means the honed bore can be smoother since it doesn't have to supply as much oil to the rings, thus reducing bore wear and oil consumption.

If you'll be operating your engine in very dusty conditions such as in desert off-road operation, use chrome faced or ceramic coated rings. Resistance to abrasive wear is considerably better than with moly rings, but the bores must be perfectly straight and round. There are disadvantages to using chrome rings: Break-in time is longer and the ends cannot be filed to increase end gap. Also, chrome rings should not be used with engines operated on propane or natural gas.

Main Bearing Bore

To avoid unequal main bearing loads from one bearing to the other, make sure the main bearing-bore axis is straight. It should also be straight if your block needs to be bored or decked. This is because the most accurate machines set up on the main bearing saddles. So, if they are not accurate, neither will the newly machined decks or cylinder bores.

To check main bearing bore alignment you'll need a precision straightedge or new main bearings and some Prussian Blue.

To make the check with a straightedge, remove the bearing caps and lay the straightedge lengthwise in the bearing saddles. With your feeler gauges, probe for any gaps between

ALIGN-BORE OR ALIGN-HONE?

A block that has its main bearings out of alignment or aren't round must be align-bored or -honed. But which one should you have done? There is a difference.

Both methods require that some material be precision ground from the main cap parting faces, or about 0.015 in./38mm, unless bore distortion or misalignment requires more removal. During both operations, the main caps must be installed and torqued to spec. This is where the similarities end. Just as the operations indicate, one method uses boring and the other honing.

Boring uses a cutting tool that can be adjusted to remove a minimum amount of material from the block side of the bearing bore. Honing removes an equal amount of material from the block as it does the bearing caps because the honing tool is centered on the existing main bearing bore. So what makes the difference? Not much if little material is removed. Otherwise it can be.

Unlike align-boring, align-honing results in the crankshaft being moved up in the block. Two undesirable things happen: Deck height—distance from the main bearing bore to the block deck(s)—moves up, reducing piston deck clearance—distance from the top of the piston to the block deck. This increases compression and reduces piston-to-valve clearance. Secondly, the crankshaft moves up in relation to the camshaft, possibly causing clearance problems if timing gears are used or slack in the timing chain of a chain-and-sprocket setup. However, honing restores a crosshatch to the bearing bores, providing a tooth that aids in preventing the bearings from spinning. Talk to your engine machinist about this one.

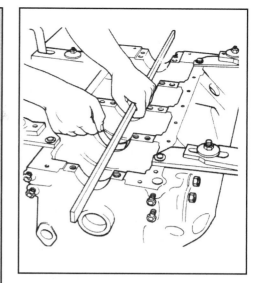

Check main bearing bore alignment with precision straightedge and feeler gauge. You shouldn't be able to slide more than a 0.001-in. (0.03mm) under the straightedge. If it's more, have block align bored. Drawing courtesy Federal Mogul.

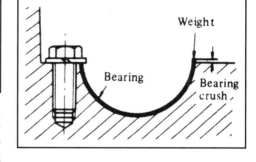

Bearing crush is amount insert extends above bearing bore parting line at one end while other is flush with split. Crush locks insert in place, preventing bearing inserts from spinning in bores. Typical crush values range from 0.0015 in. (0.04mm) for large main bearings and 0.0006 in. (0.02mm) for smaller connecting-rod bearings.

the straightedge and each bearing bore. A 0.001-inch (0.254mm) feeler gauge shouldn't fit, but refer to the manufacturer's specs to be sure of which feeler gauge to use.

Note: When making this check, the block should be set upside down on a bench or the floor. This will prevent false readings due to distortion from the block such as when it's hanging from an engine stand.

Another method of checking bearing bore alignment is used by some engine rebuilders. As described on the next page, install new main bearings, install the crankshaft—it must be within spec—and check how it turns. **Note:** Don't install the rear main seal yet. If the crankshaft spins freely, assume the journals are OK.

Prussian Blue is used with new bearings and a known good crankshaft using this bearing bore check suggested by Clevite. Just as above, install the bearings in the block. However, rather than oiling the bearings, coat the crank bearing journals with Prussian blue. With the crankshaft in place, the main caps torqued and the block on its back, rotate the crankshaft twice. Turn the block over to transfer the weight of the crank to the opposite bearing halves and again rotate the crank twice. Remove the bearing caps, lift out the crankshaft and inspect the bearings. At least 3/4 of the bluing should have transferred from the journals except adjacent to the bearing insert parting lines. This being the case, the block is OK. Otherwise, have your block align-bored or honed.

Many engine machine shops use a gauge, or a precision round bar machined 0.001 in./0.025mm smaller than the main bearing bores. The bar is installed in place of the crankshaft less the bearings and the main caps are torqued to spec. If the bar turns with the help of an extension handle, main bearing bore alignment is OK. If it doesn't, the block needs to be align bored or honed.

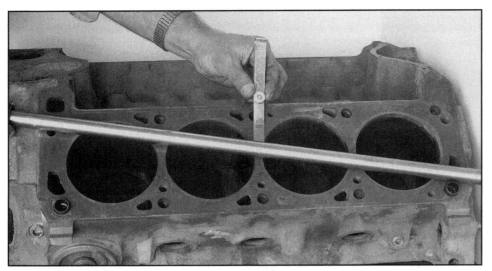

Checking deck surface with precision straightedge and feeler gauges. I suggest limiting surface irregularities to 0.004 in. (0.10mm). Check deck surface from one end to the other with bar positioned across corners and parallel to block centerline up and down deck.

Block Deck

If you find any problem with the block deck surfaces, it must be *decked*—a process where the head gasket surface is machined flat and true. This is particularly important if you found signs of head gasket leakage. If you plan on having additional machining operations done on your block, such as cylinder bore resizing or main- bearing align boring or honing, have them all done in one visit to the shop to reduce cost and save a trip.

Another more important reason to check and correct the block deck now is that many engine shops still use the old and reliable Van Norman type boring bar. This type of boring bar is mounted to and locates off of the deck surface. So if a deck is not flat or true to the main bearing bores, the remachined bores will also be off.

Checking the Deck—To check deck surface flatness, you'll need a precision straightedge and feeler gauges. The straightedge should be at least as long as the deck. Also, the deck surface(s) must be absolutely free of all gasket material and nicks and burrs. If it's not, you won't be able to make an accurate check.

To knock small nicks, burrs and small bits of gasket material off of a deck surface, run a large flat file back and forth on the deck from one end to the other. Position the file diagonally across the deck while doing this.

To check a deck, position the straightedge diagonally across both corners of the deck and lengthwise from side to side. Hold a light on the opposite side of the straightedge from which you're viewing it. If you can't see light between the straightedge and deck, don't bother checking with feeler gauges; it's OK. If you do, however, check the deck with feeler gauges at any location indicating a gap and compare it to specifications.

How flat a deck should be depends on the intended engine application and what type of gasket will be used. For instance a competition turbocharged engine with its high combustion chamber pressures and severe operating conditions should have a "perfectly" flat deck. Although achieving a perfect surface is impossible to get, you shouldn't be able to see any light passing between a precision straightedge and the deck let alone being able to fit a 0.001-in. feeler gauge between the deck and straightedge.

For most other applications, set maximum deck surface flatness between 0.002 in./0.05mm and 0.004 in./0.10mm. Also, the shorter the deck the tighter the tolerance. For example, an inline-six cylinder deck will tolerate less flatness than that of a V6.

DECKING

If you are building a race-only engine, block decking and align-boring or -honing are something you do as a part of standard engine preparation. Block decking not only ensures that a deck is flat, it also means it will have a consistent deck height for each cylinder. This ensures that the compression of each cylinder is the same providing all other work and parts are done correctly and within spec.

To deck a block, the machinist sets up the bare block in a milling machine or grinder to the main bearing bores, an important reason the block should have an accurate main bearing-bore axis. The cutters or grinding stones then cut the deck surface parallel to the main bearing-bore axis, restoring the needed accuracy to your block. Remember, if yours is a V-type engine, both cylinder banks should be decked the same amount.

When having your block decked, tell the machinist the type of head gasket you intend to use. If he's savvy—the only kind of machinist to deal with—he'll know what type of finish to use. Supply him with the gasket manufacturer's deck finish specifications as a backup.

However, make sure the above tolerance agrees with specs recommended by the car manufacturer or, preferably, the head gasket manufacturer. In addition to flatness, also note the desired finish. Some manufacturers specify a surface that is rough enough to catch your fingernail on. This is so the gasket will have something on which to grip as it is clamped between the block and head. However, because other manufacturers specify a much smoother finish, read the instructions that accompany the gasket set.

Don't remove any more material than necessary from a block deck—just enough to clean up and true the head gasket surface. The reason for this is a practical one. High octane gasoline is just about non-existent. Material removed from a block's deck surface has the same effect on compression as if it were removed from a cylinder head—compression ratio increases. In addition, piston-to-deck clearance is reduced.

Finally, if you find that one deck surface of a V-type engine is out of spec, both surfaces must be decked. This is so that both cylinder banks will end up with the same *deck height*—distance from the main bearing bore center line to the deck.

Crack Testing

You can use one or more of several crack-checking methods depending on the material being checked. Regardless of which method you use, the part being checked must be clean.

One of the better known crack checking methods is Magnafluxing. The most common but unofficial method of crack testing though is using the unassisted naked eye, making it the cheapest but least reliable method of crack checking. Other methods include using a dye penetrate and pressure testing. Let's first look at Magnafluxing.

Magnafluxing—This magnetic particle method of crack testing can only be used on ferrous-based materials, or iron and steel. To make the check, the suspected area is dusted with iron particles. Next, the area is magnetized with an electromagnet. If the poles of the magnet bridge a crack, the iron particles gather along it when the magnet is turned on, thus revealing the crack. If no crack is revealed, the magnet is turned 90° so the magnet aligns with any crack. Highlight any crack with white chalk so you'll be able to find it later.

Magnafluxing is also used to check iron or steel crankshafts, cast-iron cylinder heads, connecting rods and rod bolts.

Dye Penetrate—To use this crack-checking method, the suspect area is first sprayed with non-drying red dye penetrate. Next, the area is cleaned with solvent and allowed to dry, leaving only the dye in any cracks. For the last step, a white developer is sprayed over the same area. If a crack exists, it will show up as a bright red line as the dye in the crack bleeds through the white developer.

Another method also uses a dye, but rather than spraying the area with developer a black light is used to highlight any cracks.

CRANKSHAFT CLEANING & INSPECTION

Every part of an engine is dedicated to turning the crankshaft. This

This engine experienced catastrophic failure at high rpm as evidenced by broken connecting rod and blue bearing journal. A crankshaft must be inspected for journal damage, cracks and straightness before it's declared OK for reuse.

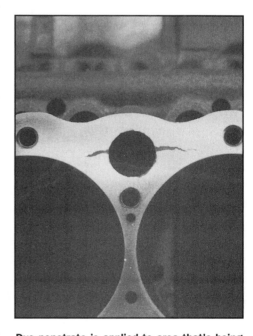

Dye penetrate is applied to area that's being checked for cracks after it's thoroughly cleaned. After cleaning again, developer is sprayed on area. Cracks show up as jagged red lines as dye bleeds through flat-white developer.

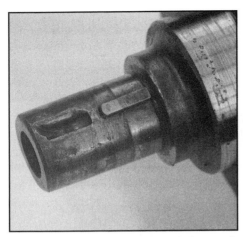

Damper bolt in nose of crankshaft loosened, resulting in keyway being hammered out of shape. Crankshaft may be repaired by welding and remachining, but replacement may be less expensive.

includes applying forces to the crankshaft, supporting it and lubricating it. Because of its importance and the loads imposed on it, a crankshaft is a very tough, high-quality component. As a result, it's rare that a crankshaft cannot be reused for a rebuild. They very rarely wear to the point of having to be replaced. What usually happens is the journals are damaged from a lack of lubrication or from poor lubrication—usually in the form of very dirty oil. This happens when the oil filter gets clogged with dirt that is bypassed.

The one thing most responsible for rendering a crankshaft useless is mechanical damage. This is usually caused by another component breaking—such as a rod bolt. The sight of such damage can make one sick to his stomach. If an engine isn't shut off immediately or if such breakage occurs at high rpm, the crank and block are usually damaged beyond repair.

Bearing Journals

The most important crankshaft inspection job involves making sure the main and rod bearing journals are round, don't vary in diameter over their lengths and are free of cracks, gouges and grooves. All imperfections can be repaired except for one: A cracked crankshaft must be replaced. There's no satisfactory method of repairing such damage.

The best way to check a crank for cracks is by Magnafluxing; a good second choice is dye penetrate. Your last option is visual inspection. Don't depend on it. The cost of Magnafluxing or using dye penetrate is minimal. Remember—a broken crank will destroy your engine.

If a bearing journal is oval or egg shaped, it is said to be *out-of-round*. This condition is more prevalent with rod journals than main bearing journals because of their loading. If a bearing-journal diameter gradually gets smaller from one end to the other, it is *tapered*.

Finally, the journals should be free from any projections raised by burrs, nicks or scoring that could damage bearings. Don't get excited by a small groove in a bearing journal. It's not going to hurt anything as long as the journal is OK in all other respects. Pay special attention to the edge of each oil hole in the rod and main journals. Sharp edges protruding above a journal surface can cut grooves in the bearings, wrecking the bearing and spoiling your rebuild. Eliminate any sharp edges by working the edge with the appropriate needle file or a very small round file.

Surface Finish—Look at the surface finish first. Any rough bearing journals must be reground—regardless of taper or out-of-roundness. Your sense of feel is better than your eyesight in judging journal smoothness, so run the end of a fingernail over the length of each journal. If it feels rough and your fingernail catches on the irregularities, it is too rough. As a final check, drag the edge of a penny over the length of each bearing journal. If it leaves a line of copper, the journal is too rough.

Regrinding—If you find a rough bearing journal, a regrind job is in order. If you're in luck, the standard 0.010, 0.020 or 0.030 in. (0.25, 0.50 or 0.75mm) regrind job will clean up all of the main or rod bearing journals. I say surfaces because if one journal is ground undersize, all rod or main journals should be ground the same. You don't want one odd-size bearing because bearings are sold in sets. The additional cost of machining all main or rod journals is slight as the major cost is in the initial setup.

Therefore, if 0.010 in. is removed from the standard bearing journal, 0.010-in. *undersize* bearings are required. Undersize refers to a variation from the standard bearing journal diameter. Example: If the standard diameter for a crankshaft's main journals is 2.450 in. and they are ground to 2.440 in., the crank is 0.010-in. undersize on the mains—0.010-in. *undersize* main bearings are used.

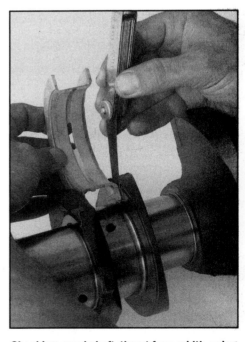

Checking crankshaft thrust-face width using thrust bearing and feeler gauges. After measuring bearing width with micrometers, add to feeler gauge thickness to determine thrust-face width. Telescopic gauge can also be used.

Thrust Faces—While you're checking bearing journals, look carefully at the crankshaft thrust faces. They're usually on the center or rear main. Be particularly attentive if your engine was backed up by a manual transmission and a stiff clutch was used. The thrust loads from the clutch being released forces the crankshaft forward with great force which must be resisted by the rear thrust face of the crank. A standard clutch can also cause thrust-face wear. If the thrust load is constant because of improper clutch-linkage adjustment—no clutch-pedal free play—excessive crankshaft thrust surface wear can result.

Check crankshaft thrust surfaces by measuring the distance between the two surfaces, or make the check by installing the crank in the block on new bearings, page 120. To make direct measurements you'll need a telescoping gauge and a 1–2-in. outside mike or you'll need a bearing insert, feeler gauges and the 1–2-in. micrometer. You'll need the factory specification for thrust width for making both measurements. If maximum end play is exceeded, you'll have to replace the crankshaft or have it reconditioned.

Out-of-Round—To check for out-of-round or tapered journals, you need an outside micrometer that brackets journal diameter. For instance a 2.500-in. journal will require a 2–3-in. mike and an idea of what you're looking for. If a bearing journal is round, as it should be, it will be described by a diameter. This diameter can be read directly from your micrometer at any location around the journal. If it is out-of-round, or egg shaped, you will be looking for the major and minor diameters of an ellipse. These can be found by measuring around the journal in several locations. Major (large) and minor (small) measurements occur about 90° from each other. The difference between the two measurements is the journal's out-of-round, or: Major Diameter - Minor Diameter = Out-of-Round.

Start by measuring the connecting-rod journals when doing your checking. They're the most likely to suffer from an out-of-round condition. Connecting rod journals are highly loaded at TDC with a decreasing load as the crank swings toward BDC. With this in mind, check each journal in at least two locations fore and aft on the journal. Typically, maximum allowable out-of-round is 0.0015 in./0.037mm. This is OK if you don't expect maximum life and performance from your engine. But if you want a top-notch engine building job, a 0.0004 in./0.01mm out-of-round limit is desirable. Remember, if you find one journal out of spec, all crankshaft rod or main bearing journals should be reground.

Taper—Bearing journal taper causes uneven bearing wear more than if the journal was out-of-round. This is because loads over the length of the bearing are not distributed

Front throw is measured for out-of-roundness. Compare measurements made 90° to each other in the same plane. I like to keep differences between these two measurements less than 0.001 in. (0.02mm) before regrinding crankshaft. Measure all bearing journals.

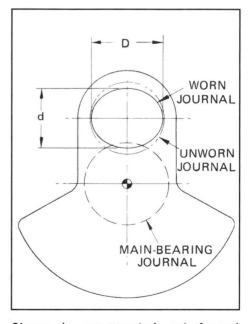

Shown is exaggerated out-of-round connecting-rod journal. Maximum wear occurs when piston is at TDC on power stroke. D - d = out-of-round.

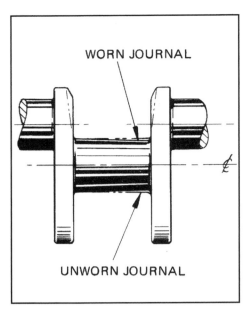

Bearing journal taper occurs when journal diameter is worn more from one end to the other. Check for taper by measuring taper along length of journal.

evenly. It concentrates its load on the less worn section, or at its largest diameter where oil clearance is least.

Taper is specified in so many thousandths of an inch per inch (hundredths of a millimeter per millimeter) of journal length. This means your micrometer readings should not exceed this amount between readings taken one inch or one millimeter apart from each other in the same plane—the mike is moved straight along the journal being measured and is not rotated.

If you're building a "new" engine, don't accept more than 0.0003 in./in./0.0070mm/mm of taper on main or rod bearing journals. However, if you're just trying to squeeze out a few more miles from your engine, 0.0012 in./in., or 0.030mm/mm of taper is maximum. Remember, if one journal is not within spec, all other rod or main journals should be reground the same undersize as well.

Crank Kits—If you need to regrind your crankshaft, consider trading it in for a crank kit. The problem may be you'll be getting someone else's reground crankshaft. If your engine doesn't have a special crank or a rare one, this shouldn't be a problem. Otherwise don't risk losing your crank and have it reworked at a machine shop.

A crank kit consists of a freshly reground or reconditioned crankshaft and new main and connecting-rod bearings. Consider taking this approach rather than having your crank reground and buying new bearings. You'll end up saving some money. Just make sure that good-quality bearings come with the crank kit before you make a final decision

Post-Grinding Checks

Check your crankshaft carefully if you've had it reground—or if it came from the kit. The factory spends a lot of time on quality control because mistakes do occur, so don't skip rechecking the crank after it's been reground.

Other than journal diameter, two problems to look for are sharp edges around the oil holes and too small or large fillet radii at both ends of each bearing journal where it blends into a throw or counterweight. All oil holes should be chamfered—edges broken—to prevent bearing damage. As for the fillet, it should go completely around each journal—where the journal meets the throw. If there is a sharp corner rather than a radius, it is a likely place for a fatigue crack to start. A radius that's too large is great for crankshaft strength, but bad news for the bearing. The large radius will cause the bearing to *edge ride.*—the fillet will extend under the end of the bearing, causing a zero clearance condition and contact the edge of the bearing.

Smooth & Clean Surfaces?—A newly reground crankshaft should've had its bearing journals polished. If you're in doubt, use the penny trick by dragging a penny the length of a journal. If it leaves a trail of copper, the journal is too rough. If it is, smooth the journals yourself rather than going through the hassle of getting it redone. Also, if your original crankshaft checks out OK in all departments, give it the same treatment.

Use narrow strips of crocus cloth to do the polishing. You can pick up crocus cloth at a hardware or auto parts store. It comes in long narrow strips. If it's wider than 1 in., tear it lengthwise to make it so.

Position your crankshaft so it won't roll around, then loop the crocus cloth completely around a journal then back on itself. If you want to get fancy, cut the strip long enough to fit a round the journal and secure the loose end with masking tape. To rotate the cloth strip, wrap it two or three times with a 3-ft. length of cotton rope. Whichever way you choose to do it, grasp each end of the cloth or rope and work evenly around the journal from one end to the

Allen Johnson polishes bearing journals with abrasive strip as crankshaft is spun between centers. Achieving smooth surface finish is objective, not material removal. You can achieve similar results with crocus cloth wrapped around journal spun with long shoe string.

Check runout with dial indicator as crankshaft is rotated on front and rear bearing inserts. Position indicator so tip doesn't interfere with oil hole. Total instrument reading should not exceed 0.0040 in. (0.10mm).

other until you end up where you started using light consistent pressure. Don't stay in one spot too long so material removal will be minimal. The object isn't to remove material, but to give the journals a clean, smooth surface.

Polish the oil seal surface, too. But rather than putting a bright polish on this surface, it must have some tooth to carry oil much in the same manner as the cross-hatch in a cylinder wall. This will help lubricate the seal, increasing its life and improving sealing.

Crankshaft Runout

Runout describes how much a crankshaft is bent. It is determined by rotating the crank between two centers and reading runout at the center main bearing journal with a dial indicator. As you rotate the crankshaft, indicator reading will change if the crank is bent.

To do a runout check, you'll need a dial indicator with a tip extension and a way to mount the indicator. A magnetic-base mount works fine. The tip extension allows you to position the indicator so a crank throw won't hit the dial assembly as you rotate the crankshaft. You'll also need two V-blocks or two main bearing inserts—old ones will do—and the engine block.

Start your runout check by setting the crank on the V-blocks or bearings in the block under the front and rear mains. Mount the indicator base to the block with the indicator positioned so the plunger is 90° to the journal. Don't forget the oil hole. The indicator should be positioned so it clears the indicator tip as you rotate the crank.

Rotate the crankshaft until you find the lowest indicated reading, then zero the dial. Rotate the crank again until indicator reading is maximum. Read runout directly. To double-check your reading, turn the crank a few more times and look for zero and maximum readings.

Total indicator reading (TIR) is found when using a dial indicator. Actual runout, or bend, is half TIR. Therefore, Runout = TIR ÷ 2. Hold runout to 0.0039 in./0.10mm, or 0.0078 in./0.20mm TIR. For best results don't let it exceed 0.0020 in./0.05mm, or 0.0040 in./0.10mm TIR. If runout readings exceed the maximum, have your crank reground or trade it in for a crank kit.

If your crank yields a runout greater than desired, try this. A crank may warp for one reason or another such as storing it incorrectly. You may be able to correct this by installing the crank in the block. Rotate the crank so its point of maximum runout is away from the block, or up with the block on its back. With only half of the center bearing insert in the cap, install the cap and torque its bolts. Leave the assembly for two days or so and remeasure runout.

If runout is now within limits, great. If not, revert to machining or a crank kit. Of if you prefer to keep the original crankshaft, find a shop that uses the hammer technique to straighten cranks. A blunt chisel placed in a fillet area and struck with a hammer does the straightening! This job is for experts, not the local blacksmith. If you live in a big city, you'll have an easier time of finding such a person. Otherwise you may have to ship your crank away for this service. Above all, don't allow anyone to use a hydraulic press to do the straightening job. It will straighten your crank, but it may also crack it.

Installation Checking Method—To do a real-world crankshaft bend check, install the crank in the block on oiled new bearings. Torque the caps to specification. If the crankshaft can be rotated freely by hand—no binding—consider it OK. Any loads induced by what runout there is will be minimal compared to the inertial and power-producing loads normally applied to the journals and bearings when the engine is running. I suggest this method of checking because crankshaft runout usually isn't a problem with a "tired" engine that just needs a rebuild. If you decide to use this approach, follow the procedure for crankshaft installation, page 122.

Journal Hardness

Some heavy-duty crankshafts have been hardened by Tuftriding or nitriding—chemical processes that introduce nitrogen into the surface at about 1000°F. Either process increases resistance to fatigue and

Although somewhat crude, one way to check crankshaft is to install it in block on oiled bearings. If it spins freely and you don't feel any movement as you try to lift nose of crankshaft up and down, it should be OK.

makes hardened journal surfaces thicker to a depth of less than 0.001 in./0.025mm, giving them a much higher wear resistance. Consider having your crankshaft hardened if maximum durability is a priority and cost is not a major consideration. Check with your machinist on this one.

To check a crankshaft for surface hardening, file a non-bearing surface such as the nose of the crank. If it hasn't been hardened, the file should cut easily. If your crank has undersize journals, it's possible that it was turned down and not rehardened. If you suspect this, carefully file a journal immediately adjacent to an oil hole with the edge of a small file. Again, if it cuts easily, the journal surface isn't hardened.

Final Cleaning

Now that crankshaft reconditioning is complete, make sure it's clean. Even if you had the crank hot-tanked, clean out those oil passages. Make sure they are perfectly clean. Do this by pulling a solvent or lacquer-thinner soaked piece of cloth through the oil holes with a wire.

A nylon coffee pot brush or a small gun-bore brush works great for cleaning crankshaft oil passages. If you used one to clean the block oil galleries, it will work fine. Consider this a necessary step because even a hot tank or jet spray won't get all the crud off of or out of a part. To finish up, use a stiff bristle brush and solvent on the throws and counterweights if they look dirty. Finish by wiping with paper towels.

Treat the crankshaft with a coating of rust-preventative oil. Make sure the bearing journals are well covered. Unless it's to be used with a manual transmission, you can slip your engine-ready crank into a plastic trash bag and store it out of the way. If a manual transmission is to be used, you have one more job.

Pilot Bushing or Bearing

Crankshafts used with a manual transmission have a pilot bushing or bearing installed in the center of the flywheel-mounting flange. This is used to support the nose of the transmission clutch shaft, or input shaft.

Bushings are solid oil-impregnated sintered-bronze discs with a hole in the center. Bearings are usually the needle type, however some installations use sealed ball bearings. If your crank was used with a manual transmission, it'll have a bushing or bearing; if it's a new or reconditioned crankshaft, it won't have either.

Crankshafts used with automatic transmissions won't have a bushing or bearing. So make sure there's neither if you're going to use an automatic behind your engine. Remove the bushing or bearing. Continue reading to determine how to do this.

If your crank is equipped with a bushing, visually check its bore for damage. If you have an old transmission input shaft, insert it into the bushing and check for side play. If there is noticeable movement or you have the least doubt, replace the bushing. The same goes for a bearing. If you have any doubts about its condition or it is a high-mileage item, replace it.

Bushings/bearings can be removed with a slide hammer. The slide hammer jaws reach inside the bore and hook against the back of the bushing/bearing. A heavy slide is impacted against the slide hammer handle to knock the bushing/bearing free from its bore.

Solid bushings can be removed with the hydraulic method if you don't have a slide hammer. Start by filling the bushing bore and cavity in the crank behind the bushing with grease. Poke it in real good, making sure it's all grease and no air. Note how deep the bushing is installed in the crankshaft.

Insert a snug-fitting dowel or the nose of an old input shaft into the bushing bore, then drive it in with a heavy hammer. This will pressurize the grease and push out the bushing. After the bushing is out, remove all of the grease except for a light film on the sides of the bore.

Bushing Installation—To install the new bushing, find a short round section of soft metal such as aluminum or brass that has a slightly smaller OD than the bushing. Use this as a driver between your hammer and the bushing. Or, if you have a brass hammer, use a deep socket as the driver if you don't mind hammering against it.

Position the bushing squarely over its bore in the crankshaft and begin driving it in. Get it started with light blows, making sure it doesn't get cocked in the bore. If it does, remove the bushing and start over. Once started straight, drive the bushing in to

If used with manual transmissions, install pilot bearing. Drive pilot bushing or bearing in with hammer and hollow punch. If bearing is used, make sure punch seats against outside race only.

the same depth as was the old bushing.

Bearing Installation—Install a pilot bearing just as you would a bushing. If it's a needle bearing, it won't have an inner race but it will have a cage to retain the needles. In the case of a ball bearing, it'll have inner race, an outer race and a seal. In both cases, make sure the driver bears against the outer bearing race only so you don't destroy the bearing assembly. Also, make sure the seal is installed so it will be toward the transmission, not the engine.

CAMSHAFT & LIFTERS

A camshaft lobe controls valve opening: how soon, how fast, how far and how long the valve opens. Although it rotates at only half crankshaft speed, a camshaft used with conventional lifters—flat tappets—is more prone to wear. Camshafts used with roller lifters are more durable.

Camshaft lobes are subject to extremely high contact pressures at the lifters. This is particularly true with high-performance or racing cams. These "see" even higher loads because of high lifts, solid lifters, high spring pressures and high operating speeds.

Unlike a crankshaft, once a cam lobe or lifter starts wearing, it's not long before the lobe and lifter are "gone." This is especially true with conventional lifters where there's a wiping action of the lobe against the lifter foot. So give your old camshaft and lifters a very close look before deciding to reuse them.

Two hard and fast rules apply to camshaft and lifter usage:

1. When reusing a camshaft and its lifters, the lifters must be installed in their original order. What this says is the lifters must be installed on the same lobes, or in the same lifter bores. Caution: In the case of a roller lifter, the lifters must be installed so they roll in the same direction as they did originally.

2. Never install used lifters on a new camshaft. If you think the old lifters look good enough to reuse, think again. They can't be good enough to reuse. However, it's OK to use new lifters on a used camshaft providing the lobes are in good shape.

In addition to the above rules, I also recommend that you only install a used but good cam and its lifters in the engine in which they were originally installed. In another engine, the lifter bores will not be in the exact relationships to their cam lobes as they were in the original engine. This would have the same effect as mixing up the lifters: wiped cam lobes.

Cam Lobe & Lifter Terms

Before you inspect your camshaft, you should have a basic

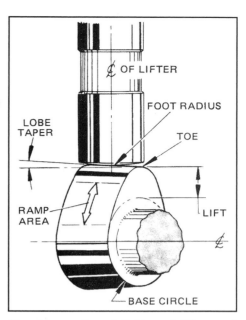

Visual inspection of flat-tappet lifters and cam lobes will tell you whether they should be replaced. If wear pattern is full width of cam lobe over of toe area and lifter foot is worn flat or concave, camshaft and lifters need replacing.

understanding of camshaft and lifter terminology. You should also understand a little about cam and lifter design and the reasons for these designs.

As shown in the sketch, the two major sections of a cam lobe are the toe and the heel. The *toe* is the point on the lobe contacted by the lifter when the valve is at maximum lift; the *heel* is directly opposite the toe, or the lobe-to-lifter contact point when the valve is closed. The heel is described by a radius—the *base circle*—swung off the axis of the camshaft. The transition areas of the lobe during valve opening and closing between the toe and heel are called the *opening ramp* and *closing ramp*, respectively.

Flat-Tappet Lifters—These are not really flat—the portion that runs against the cam lobe is actually convex. The lifter-to-lobe contact area, or foot, has a large spherical radius. Additionally, the lobe of a flat-tappet cam is not square as viewed from the side, but is tapered. This taper, the spherical lifter foot and an

59

Unlike flat-tappet cam that uses tapered lobes, roller cam-lobe surfaces are parallel with centerline of cam. Another difference: roller cam is machined from steel billet rather than cast iron.

Roller lifter pair and tie bar: Check roller surface for wear and roller as you would a bearing. Rotate roller and feel for roughness. If you see or feel any damage, replace lifter.

off-center lobe-to-lifter contact combine to minimize lobe and lifter wear by rotating the lifter in its bore. This rotation constantly exposes a new lifter wipe area to the lobe; the spherical lifter foot ensures minimum contact area between the lobe and lifter.

Roller Lifters—These lifters are true to their name. Rather than having a "flat" foot, a roller that rotates on needle bearings is placed at the lifter foot to reduce friction and wear. This also allows for a more radical lobe design, thus allowing the engine to produce more horsepower than would be possible with a flat-tappet cam.

Note: Special flat-tappet cams called mushroom tappets are available. They are capable of radical lobe designs similar to those used with roller lifters, however they still retain the high friction and wear of the conventional flat-tappet lifter.

A roller lifter is longer and it must be centered on the cam lobe. Also, the lobe is not tapered, but square. To prevent catastrophic lifter wear from the roller "skidding" sideways on the lobe, the lifter is prevented from rotating in its bore by a link, bar or guide plate attached to the adjacent lifter or the block.

Camshaft Inspection

If you insist on trying to reuse the old camshaft, let's see if it and the lifters are serviceable. Every cam/lifter inspection starts with the cam. If it's bad, you must replace both the cam and lifters regardless of lifter condition. Remember the second rule; never install used lifters on a new camshaft.

If you checked valve lift with a dial indicator, there is no need to check lobe lift now. Check back to your notes to see which if any lobes are worn. Remember, to be accurate you should've measured with a dial indicator at the end of a pushrod. If you didn't measure the cam lobes before, do it now. You'll need a pair of dial calipers or a micrometer to measure lobe lift.

Measuring Lobes—Measuring the lobes is a comparison check, one lobe versus another. Unlike the check in the diagnosis chapter, the purpose here is to compare relative lobe wear, not lift. Because cam lobes wear at a rapid rate once they start wearing, you should have little trouble pinpointing a bad lobe.

Note the wear pattern on each lobe, particularly at the toe. In the case of a flat-tappet cam, if wear has extended from one side of the toe to the other and well down on both ramp areas, the cam has seen better days and

Lobe lift being measured with vernier calipers. Measure across base circle, then across heel and toe. Difference is gross lobe lift. Measurement is not accurate with high-performance cams as ramp areas extend too far around lobe, thus affecting reading.

Most accurate way of measuring lobe lift is with dial indicator on lobe. Indicator is zeroed when lifter is on lobe base circle, then read as lobe lift when needle shows highest indicated reading as cam is rotated. Note link tying roller lifters together.

Camshaft bearing journals are getting same polishing treatment as crankshaft. Abrasive belt runs against journal as cam rotates between centers.

should be replaced. The same goes for pitting. Any pitting on a lobe surface, flat tappet or roller indicates metal loss which will probably show up on its mating lifter. A full-width wear pattern or pitting on a lobe surface indicates the camshaft and lifters should be replaced. If the cam appears to be OK, measure it. Use dial calipers or a micrometer.

Take the first measurement across the lobe, halfway between its heel and toe; this is the minor dimension. (On a mild camshaft it will also be the base circle diameter.) Then measure the major dimension on the same side from the heel to the toe as shown. Both measurements should be made at the points of maximum visible wear and the same distance from the same edge of the lobe. Now subtract the minor dimension from the major dimension to get approximate lobe lift.

You can't depend on these measurements to reflect lobe lift accurately, particularly with a high performance cam. This is primarily because you'll be measuring across the ends of the clearance ramps—transitions between the base circle and the opening and closing ramps. The reading you get will be greater than the base circle diameter. As a result, lobe lift will appear to be less than actual.

Measure all lobes, recording your results as you go. In the typical pushrod engine, there will be two lobes for each cylinder, one for the intake valve and one for the exhaust. Compare your findings by placing lobe-lift figures for the intake lobes in one column and exhaust figures in another. Lifts for both intake and exhaust should be within 0.005 in./0.13mm of each other. If there is one bad lobe, the cam and lifters must be replaced.

Cam Bearing Journals—Camshafts are more durable than crankshafts in one respect; their journals seldom wear out. Don't take this as an excuse not to check them. All should measure within the limit set by the manufacturer.

As with the crank, there are runout and out-of-round limits for the camshaft. However, if it turned easily in the old bearings, it'll be OK as is.

To check the cam out of the block, support it with two V-blocks, one under each end journal and set up a dial indicator on it center bearing journal. Or support it between centers, such as in a lathe. Indicate the center bearing journals as you did with the crankshaft. Runout should not exceed 0.0015 in. in most applications. Check out-of-round with dial calipers or a micrometer. Limit out-of-round to 0.0005 in./0.013mm.

The final check will be to install your cam in the block and spin it in well-oiled bearings. Unless the cam shows play or binds as you turn it, there is no problem.

As far as lifters are concerned, I highly recommend that you buy new ones, even if you are using the old cam.

PISTONS & CONNECTING RODS

If you are having your block bored, trash your old pistons. However, if bore wear is acceptable you can get by with merely honing. Just make

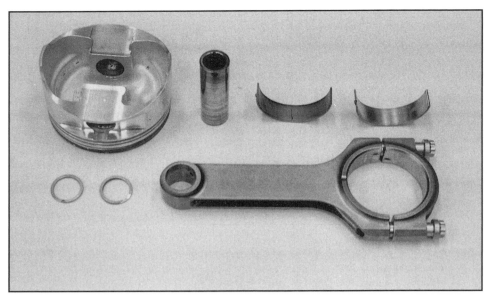
Disassembled connecting rod and piston. Full-floating wrist pin is held in place with Spiralox-type retainers. Pressed pins are held in place by an interference fit between small end of rod and pin.

Sometimes broken skirt is evident, sometimes not. It may just be collapsed. Either condition is reason to replace piston.

sure the pistons are usable.

Pistons, like the timing chain, sprockets, camshaft and lifters, see a lot of abuse and wear; but they don't wear as badly. Although an engine may have been well maintained, if it's considered to be a high-mileage engine, most mechanics or machinists opt for "new" bores that can only be achieved by boring.

Engine builders reason correctly that fresh bores and new matching pistons will give them as-good-as or better-than-original life and performance. After all, used pistons are just that—used. Some of their working life has already been spent. That brings up the final consideration in the piston/bore relationship.

Even though you may have found the cylinders don't need reboring, you may also have found that, after inspecting them, the pistons should be replaced. If this proves to be the case, the reasonable thing to do is rebore and install oversize pistons—most of the time. An exception would be where you want to retain the same bore size or the bore is already to its maximum oversize. You simply want to retain the engine's original displacement, the block is of the rare variety or the engine will be used in competition where maximum displacement would be exceeded if you bored.

New pistons are the major expense in a rebore, so you won't have to spend much more to get straight bores. You'll also get a slight displacement increase. This amounts to about a 1 to 2-cid increase for every 0.010-in./0.25mm increase in bore size for an eight-cylinder engine. For example, a V8 with a 4-in. bore and a 4-in. stroke displacing 402 cid will displace 408 cid with a 0.030-in. bore increase, only 6 more cubic inches! The 1.5% displacement increase doesn't amount to much if it's a displacement increase you're after.

A final consideration in your rebore-or-not-to-rebore decision making: The thinner cylinder walls that result from enlarged bores will result in more heat being "wasted" to the coolant. The consequence of this may be a hotter running engine—at least as far as the coolant is concerned. The bottom line is you may have to go to a bigger radiator to prevent overheating.

Inspect Pistons

If you are not going to rebore, inspect the pistons before removing them from the rods. If they are OK, you can rebuild the engine without disassembling them except in two cases; the pistons and rods should be separated if full-floating pins are used or your engine has been used in very heavy-duty applications such as racing. This will allow closer inspection of the pin bores for wear and piston-pin bosses for cracks. Just be careful to keep all of the pieces together. However, for the standard rebuild, don't bother with removing the rods from the pistons. It is one less job, and will greatly speed the rebuild. It may also mean one less trip to the machine shop in the case of having to press out the pins. It also means one less bill to pay.

Ring Removal—Remove the old rings. There's an easy and a hard way to remove piston rings: either with a ring expander—the easy way—or by hand. Doing this job by hand is tiring, is hard on your thumbs and increases the changes of scratching a piston with the ring ends. However, using a plier-looking ring expander grips the ring ends to expand, or spread, the ring, making it easy to lift the ring over the piston with much less chance of scratching the piston.

Both methods are easier when the rod-and-piston assembly is held steady by something other than your own hands. A vise is ideal, but be very careful not to mark or bend the rod. Lightly clamp the rod between two pieces of wood in vise jaws. Soft lead, brass or aluminum vise jaws will also work.

Clamp on the edges of the rod beam and position the piston so the bottom edges of its skirt rest on the vise jaws or wood blocks. This will keep the piston from flopping back and forth as you work with it.

If you don't have a vise, clamp the rod and piston to a bench with a C-clamp. Use a wood block between the clamp and rod to prevent damage. If the bench is metal, put a another wood block between the bench and rod. Lightly clamp the rod to the edge of the bench with the piston resting against the bench top.

Piston Damage—Obvious damage to a piston results when a valve drops into a cylinder, damaging the top, or dome, or one side of the piston—a skirt—breaks off. You'll have to look closer for other types of damage. Scuffing, scoring, ring-land damage and skirt collapse can be found through visual inspection.

Scuffing describes abrasion damage to the piston skirt. This is usually caused by insufficient clearance or lubrication between the bore and piston. Scoring is similar to scuffing, but is limited to deep grooves in a skirt. Both types of damage are caused by lack of lubrication, severe engine overheating—causing insufficient piston-to-bore clearance—or a bent connecting rod.

The common denominator in all three is excessive pressure and temperature between the piston skirts and cylinder wall. The most common cause of scuffed and collapsed skirts is overheating caused by a stuck thermostat, blocked water passage, broken fan belt or low coolant.

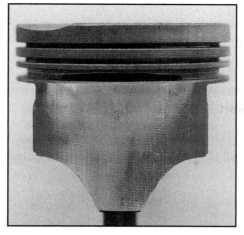

Skewed wear pattern on skirt indicates connecting rod is bent or twisted. Rod should be checked in fixture.

Wear pattern on skirt is straight and symmetrical. This indicates connecting rod is straight.

If a rod is bent to the side or twisted, it will cause the piston skirt to wear or scuff unevenly from top to bottom. A little variation in the skirt wear pattern is normal, but when a piston skirt is scuffed or worn other than straight up and down in the center of a skirt, you can be sure the rod is bent or twisted.

Checking connecting rod alignment is a normal part of machine shop routine, but it never hurts to let them know about your findings. You can't check the rods accurately for straightness because you don't have the equipment—but you may have the evidence that they need checking.

Piston Skirt Diameter—Measuring piston skirt diameter is not so much to determine wear as it is to check for collapsed skirts. Before measuring and judging the results, you should fully understand what piston skirt collapse means.

Pistons are widest at their skirts, and they get wider from the oil ring groove to the bottom of the skirts. The reason for this is heat expansion control. As a piston heats up from cold start to operating temperature, the aluminum expands. As it expands, the desire is for skirt width to be the same from top to bottom. The part of the piston closest to the heat source—the combustion chamber—and farthest from crankcase oil cooling, expands the most.

In normal service, piston and cylinder walls warm to their operating temperatures and the piston skirts expand to a given size and shape. But, if the engine overheats, the piston skirts continue to expand until all piston-to-bore clearance is gone; the skirt is forced against the cylinder wall and begins to scuff. This scuffing action increases piston heat even more, but the cylinder wall doesn't permit further expansion. The skirts are then overstressed, or overloaded, and one bends or breaks off. After the engine cools, the bent piston skirts contract to a smaller diameter than before—they are said to have collapsed. The result will be greater-than-acceptable bore clearance, causing slap during warm-up and, maybe, after the engine has reached operating temperature.

If you used pistons with collapsed or missing skirts—heaven forbid—they would be noisy. Additionally, the oil ring grooves may have been

Measure piston diameter 90° to and level with bottom of wrist-pin bore. Measurement across bottom of skirt should be slightly more, or about 0.0005 in. (0.01mm)

distorted as well, resulting in poor oil control.

Heavy scuffing is a common indicator of collapsed piston skirts. To confirm this, measure the piston in two places and compare the readings. First, measure the skirt level with and 90° to the piston pin. Move straight down the skirt to the bottom and measure its diameter again. Generally, cast pistons should measure 0.0005-in./0.013mm wider at the bottom of the skirt; forged pistons should measure 0.0007-in./0.018mm wider. If they are the same or are narrower, chances are one or both skirts have collapsed. Junk the piston.

Inspect Domes—The piston is first to suffer from excess heat caused by preignition or a lean air/fuel charge. Such damage removes aluminum from the dome and deposits it on the combustion chamber, valves and spark plug or blows it out the exhaust port. This leaves the dome with a spongy or porous appearance to which carbon attaches.

Clean the piston domes so you can inspect them. If the carbon buildup is heavy, remove it with a dull screwdriver or old compression ring. Make sure the screwdriver is old and worn out. One with a sharp blade will scratch the piston easily. Obviously, you must not use a sharp edge scraping tool on the pistons such as a chisel or gasket scraper.

Follow the screwdriver cleaning with a wire brush or wheel. Be extra careful to keep the bristles away from the ring grooves. Light scratches on top of a piston won't hurt, but they will cause trouble at the grooves. Scratches here will provide leak paths between the rings and ring lands.

Piston Pin Bore Wear—Although some piston pins, or wrist pins, are retained with a lock bolt at the small end of the rod, the two most common methods of piston pin retention are with an interference fit—commonly called *pressed pins*—and retaining rings—used with full-floating pins. The main difference between the two is the pressed pin has an interference fit between the pin and small end of the rod—the pin is bigger than the bore in the rod. With the full-floating pin, the pin is free to rotate in the rod because there's clearance between the pin and rod. Retaining rings are used at each end of the pin to keep it from sliding out of the piston bore and into the cylinder wall.

How you check pin bore wear differs with how the pin is retained. To measure a pressed pin and its bore, start by soaking each piston in solvent or carburetor cleaner. Work the rod back-and-forth while the pin is submerged. This will remove the oil cushion from between the pin and its bore. Place the piston-and-rod assembly upside down on a bench, grasping the piston with one hand and the big end of the rod with the other. Hold the piston with the end of your thumb across one end of the pin and a finger across the other. Now try to rock and twist the rod without moving the piston as shown in the photo. Rock the rod in the direction of the pin centerline. Next, twist it in a circle concentric with the piston circumference. If wear is excessive you'll be able to feel pin movement at your fingertips.

Again, if you feel any movement, the piston pin bore is worn. Chances are the remaining piston pin bores will be worn too if the first one you checked is worn. If this is the case, the answer to correcting the problem is to buy new pistons. To be sure, though, check the remaining pistons.

Check piston pin bore wear by grasping piston and rod as shown and rocking rod from side to side, then twisting rod. Place finger and thumb over end of pin. If you feel any movement at pin, piston should be replaced. Pin-to-bore clearance should be about 0.0005 in. (0.01mm).

Pistons and connecting rods with floating pins are easy to disassemble, so take them apart and measure the pin, piston and rod with a micrometer and a telescoping gauge. Use a dial bore gauge if you're fortunate enough to have one. Use retaining ring pliers to remove the retaining rings at each end of the piston pin. Slide the pin from the piston and rod. Next, mike the pin at each end where it installs against the piston. Record your measurements. Now, measure the piston pin bore ID at both ends and record these figures.

Subtract the pin diameter from pin bore ID to get pin-to-bore clearance. Floating pins should have 0.0004–0.0008-in./0.010–0.020mm pin-to-piston clearance. If this clearance exceeds 0.0008 in./0.020mm, replace the piston. Measure the pin in the center and the small end of the rod. Subtract these two figures to get pin to-rod clearance. It should be 0.0005–0.0007 in./0.013–0.018mm.

In case you are wondering why pin clearance at the piston is "wider" than at the rod, the thermal expansion of aluminum is higher than it is for steel. The result: pin clearance at the piston changes more than at the small end of the rod with temperatures changes.

Caution: Check the car or piston manufacturers' specs for piston pin clearances. Although the clearances I've given are accepted industry practice, deviations may occur.

If you've disassembled your pressed-pin rod-and-piston assemblies, measure them. Piston-to-pin clearance should be 0.0003–0.0005 in./0.008–0.012mm, but this clearance can safely go to 0.001 in./0.03mm, the upper tolerance limit. At the rod, there should be a 0.0008–0.0012-in./0.020–0.030mm interference fit with the pin—the pin is larger than the small end of the rod. Going to the high side with this limit is best just to ensure maximum pin retention. Too much is not good, though, because if the pins are pressed in, they may gall and cause rapid piston-pin-bore wear.

Checking for Cracks—If you are reusing forged pistons, carefully inspect their pin bores and bosses. These weak areas are prone to cracking, which is followed by complete failure. Unaided visual inspection is not very good at spotting cracks. However, two types of non-ferrous crack checking can be used: Spotcheck (a dye test kit) and Zyglo.

After the area being checked is thoroughly cleaned and dried, it is sprayed with a liquid penetrate and allowed to soak in for a few minutes. The surface is then washed. At this point, using Spotcheck and Zyglo differ. With Spotcheck, the area is sprayed with a developer—a white powder. A surface crack will show up as a jagged red line as the dye in the crack bleeds through the developer. For Zyglo, the part is viewed under a black light. Cracks will show up like a neon light. If you find any cracks, replace the piston.

Ring Grooves—Ring groove condition is vital to ring functioning.

Adjust ring groove cleaner to groove width and diameter, then run tool around groove to remove carbon deposits. Be careful not to remove metal.

Rings must have a good seat on which to seal, so inspect the ring grooves carefully.

To do this, the grooves must be clean. To accomplish this, you'll need a ring groove cleaning tool. Two different types can be used; one's free and the other is cheap. The free tool is a broken ring. The cheap one is specially designed to clean ring grooves—it's a ring groove cleaner!

If you don't want to spend the few bucks for a "one-time-use" tool, use an old compression ring. One that came off your pistons is best because you know they're the right size. Start by breaking the ring in half and grind or file one end. Break the sharp corners on all the cutting edges. This will keep you from removing metal from the ring lands. These surfaces must remain as tight and as damage free as possible for good ring sealing.

The trouble with using a broken ring is that the ring can scratch and gouge the ring groove if you don't take great care while you're cleaning them. Once a ring groove is damaged, the ring cannot adequately seal between the ring and piston.

Because it is faster and there is less chance of damaging the piston, I prefer to clean the grooves with a ring-groove cleaner. As shown, a ring-groove cleaner pivots in the groove as it is rotated. It does an excellent cleaning job without much danger of damaging the grooves.

Measuring Ring Grooves—The easy way of measuring ring groove width or ring side clearance is with a new ring and feeler gauges. New rings present a problem, though. To get one new ring, you'll have to buy a complete set of new ones just to find out that your old pistons may be junk. Assuming you'd do the logical thing and rebore your engine and install oversize pistons and rings, you'd be stuck with a new set of rings that you had no use for. To avoid this, use an old ring and compensate for its wear.

Typical top compression-ring widths are 1/16 in. (0.0625 in./1.59mm, 5/64 in. (0.0781 in./1.98mm) and 3/32 in. (0.0937 in./2.38mm. Ring-to-groove side clearance is best at about 0.0025 in./0.06mm, but can be in the 0.0017–0.0032-in./0.04–0.008mm range. To simplify the checking process, consider 0.003 in./0.08mm as maximum ring side clearance. Correct ring groove width will then be the sum of the ring width and its side clearance.

To measure ring groove width, mike the thickness of one of the old compression rings. Subtract this from the original thickness measurement of that ring. Add this difference to the ring side clearance specified. This will be your checking clearance.

Now insert the edge of the ring in the groove and gauge the distance between it and the groove with your feeler gauges. You have the checking clearance right away, but you must check all the way around the groove. When you find the feeler gauge that

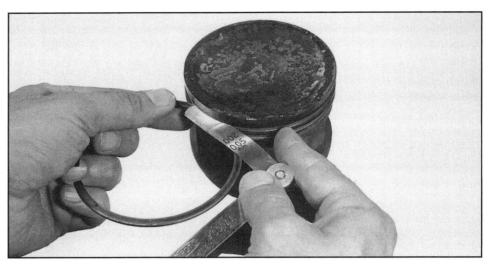

Check ring-groove width with ring and feeler gauge. Ideal side clearance is 0.0025 in. (0.06mm) and absolute maximum is 0.006 in. (0.15mm).

fits snugly—not tight or loose—subtract the measurement from the checking clearance. The difference is piston ring side clearance—the dimension used to determine if the piston ring grooves are serviceable.

Now slide the ring and feeler gauge around the piston in the ring groove and you can easily determine whether the ring lands are bent or worn excessively. The feeler gauge will be loose or tight where the ring land is bent and loose where it is worn.

Example: Let's say you have 1/16-in. (0.0625 in./1.59mm) compression rings. You mike one of the old rings and get 0.0615 in./1.56mm, or 0.001 in./0.02mm of wear—about normal. Because 0.003 in./0.07mm is the ring groove checking clearance with a new ring, add 0.001 in./0.02mm to 0.003 in./0.07mm when using the old ring; 0.004 in./0.10mm is your new ring-to-groove checking clearance.

After measuring the ring groove with the ring and feeler gauges, you find a 0.003-in./0.07mm feeler gauge fits best. The ring groove is OK because checking clearance is under 0.004 in./0.10mm.

If the first and second compression ring grooves use the same width rings, use the same ring to check all pistons and grooves. Otherwise, use the same procedure with the correct ring. Start with the top groove. It should be the worst. Don't skip checking both compression ring grooves, though.

Oil Ring Grooves—Because the oil rings and their grooves are drenched in oil and lightly loaded—compared to compression rings—they don't wear much. Therefore, if the compression ring grooves are OK, visually inspect the oil ring groove in each piston for obvious damage and let them go if you see no problem. To be absolutely sure all is OK, go ahead and measure them.

Oil rings use much wider grooves than compression rings. You can't use oil rings, so stack two compression rings together and use them in combination with your feeler gauges.

To establish an oil ring groove checking clearance, mike the two compression rings stacked together. Subtract this figure from the oil-ring width, which is typically 3/16 in. (0.188 in., or 4.76mm), plus a maximum side clearance of 0.006 in./0.15mm, or 0.194 in./4.93mm. The difference is checking clearance, or the feeler gauge you'll use with the two rings stacked together.

Example: If you stacked two rings

Placed in checking fixture, piston is rocked back and forth while indicator mark is observed. If it moves, rod is twisted or bent. If bend or twist is minor, rod can be straightened and reused.

together and got a combined thickness of 0.155 in./3.94mm, this figure subtracted from 0.194 in./4.93mm gives 0.039 in./0.99mm—the thickest feeler gauge you should be able to get between the two compression rings and in oil groove. If you encounter the unlikely situation that an oil ring groove is worn, junk the piston.

Inspect Connecting Rods

To accurately inspect the connecting rods, you'll need special tools. They are so expensive that you probably can't justify purchasing such equipment unless you own your own pro shop. This is a job to farm out unless the pistons and bearings tell you otherwise

For example, if piston and bearing wear is even and your engine has not been abused, you may be able to skip checking the rods. On the other hand, connecting rods are the most critical bottom-end part, particularly if your engine is being built for heavy-duty use or racing. So have the rods checked by an engine machinist even though the pistons and bearings don't exhibit unusual wear patterns.

The major areas to check on connecting rods are for bending or misalignment, out-of-round bearing bores or bores damaged by a spun bearing, rod bolt condition, and, in the case of full-floating wrist pins, wrist pin bores that are too large or damaged from a failed bushing.

If you are building an engine that will be used for racing, you must have the rods Magnafluxed. This is especially true of the rod bolts.

Bent or Twisted—Take a close look at the piston and rod bearing wear patterns. The wear pattern on the piston should be vertical and symmetrical across both thrust surfaces. Look at the major thrust surface—the side opposite crank rotation. One side of the wear pattern should be a mirror image of the other. However, if the wear pattern is skewed, page 63, chances are the connecting rod is bent or twisted.

As for the rod bearings, they should also show even wear from side to side. A bent rod places an uneven load on its bearing and journal. The top insert will be worn on one side and the bottom insert will be worn on the other. A similar wear pattern will be caused by a tapered bearing journal, however the wear will be on the same side of each bearing insert.

For example, if a rod is bent, the upper insert will be worn on the rear portion while the lower insert will be worn on the front, or vice versa. Double check the crank if you find such a wear pattern.

If you suspect that a rod is bent or twisted, take it to an engine machine shop and have it checked and straightened. They'll have a fixture for checking connecting rod alignment. This usually can be done with the piston on or off.

If the rod is bent 15° or less for a standard rebuild, it's OK to straighten and reuse it; 5° if the rod is for high-performance use. Replace the rod if it exceeds one of these figures.

Inspect Rod Bolts—Unless hollow alignment dowels are used with bolts that thread into the connecting rod, the bolts align the cap to the rod. Therefore, if you will be installing new bolts, they must be installed before inspecting the big end and resizing it. Also, if new bolts are installed or you remove and reinstall the original bolts, you must resize the connecting rods. In both cases, this is because cap-to-rod relationship changes. Keeping these points in mind, inspect the rod bolts now.

If your engine won't see heavy-duty or racing use, you can probably get by with visually inspecting the rod bolts for cracks. You'll need a bright light and a magnifying glass for this. But visual inspection can't be done well without removing the bolts. So bite the bullet and remove the rod bolts and have them Magnafluxed.

Remove pressed-in rod bolts by first clamping the rod with the bolts pointing up in a rod vise or bench vise between two blocks of wood. A rod vise has smooth, parallel jaws so it won't scar or distort the rod or cap.

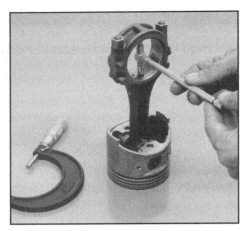

Cap is installed and bolts or nuts are torqued to spec in rod vise prior to measuring and resizing. Telescopic gauge is set in line with rod beam, the direction that cap distortion occurs to give an oversize condition.

The wood blocks will prevent rod damage when a bench vise is used

To protect the rod bolts, install a nut on each bolt so it's flush with the end of the thread. Drive out the bolts with a soft hammer or a brass punch between the hammer and bolt. After inspecting the bolts, reinstall them with a press or draw them in by installing the cap and tightening the nut. Again, correctly secure the rod in a rod vise.

Rod Bearing Bores—Both big- and small-end bores—particularly if full-floating pins are used—should be checked for size and out-of-round. You should've already inspected the wrist-pin bores during a preliminary check, so now concentrate on the big end.

To obtain the needed bearing crush, the big end of a rod must have a slightly smaller bore than the OD of the bearing insert. This slight undersize causes the bearings to crush, locking them in place and forcing the insert into intimate contact with the rod bore. They must also be round so the bearing will be round.

Bearings need a tight fit for two reasons: to lock them in place and to provide maximum bearing-to-rod heat transfer. Yes, it's bearing crush that keeps a bearing from spinning, not those little tangs, or tabs, at one edge of each insert. Contrary to popular opinion, bearing tangs are there only for positioning bearing inserts in their bores during assembly. Once the bearings are in place, the tangs serve no useful purpose.

Bearing crush can be excessive. If the bore is too small, the bearing-insert parting lines will be forced inboard toward the bearing journal. This shows up as excessive wear at the bearing parting lines.

If the bearing bore is too big, there will be insufficient crush. This will cause the bearing to overheat. Additionally, the bearing will move in its bore, possibly to the point of spinning. The danger here is it may ruin the rod and the crankshaft.

If a full-floating wrist pin is used and the wrist pin bore is too large, the wrist pin will knock. This will beat out the bushing and possibly destroy the wrist pin and rod. If it is too small, the wrist pin will seize and possibly score the piston pin bore and knock out the retaining rings. Additional damage to the cylinder bore will likely result.

The best way to check your rods and correct them if a problem is found is to take them to an engine machine shop. They'll have a connecting rod reconditioning machine where both ends of each rod are checked for size and roundness. If a problem is found, the connecting rod is resized by honing. The shop will also have two other special pieces of equipment: a rod vise for holding the rods when the cap is torqued in place and a special grinder for removing material from the cap parting face prior to resizing.

To measure your rods "on site," you'll need an inside micrometer, a dial bore gauge or a telescoping gauge. You'll also need an outside mike when using a dial bore gauge or a telescoping gauge. Before checking the big end, install the cap on the rod. If don't have a rod vise, clamp the rod in a vise between two blocks of wood. Install the cap on the rod as you would when installing it in the engine and torque the nuts or bolts to spec.

Big End: At the big end, mike the bore in line with the beam then at several different angles to check for out-of-round. If you are using a Sunnen style indicator, rotate the rod to check for size and out-of-round. Write down the measurements for each rod. Typically, out-of-round and size is limited to 0.0004 in./0.010mm, or +0.0002 in./0.005mm.

If the big end of the rod is out-of-round or oversize, the mating surfaces of the cap must be precision ground. This is to reduce the big end diameter as measured in line with the rod beam. Material can then be removed by honing to bring the bore within spec. All honing on a connecting rod should be done on a precision hone such as the one pictured so big- and small-end bores will be correctly sized and aligned.

A final word about the big end: It's a good idea to replace the rod-bolt nuts, especially if they were removed with an impact wrench. Buy new nuts now so you'll have them for engine reassembly.

Small End: When sizing the small end full-floating pin, there should be 0.0001—0.0005 in./0.002—0.012mm pin-to-bore clearance for a standard rebuild. For all-out racing, full-floating pin clearance is increased to 0.0008–0.0010 in./0.020–0.025mm. For pressed pins, there should be a 0.0008–0.0012-in./0.020–0.030mm interference, or press, fit between the pin and small-end bore.

If your engine uses pressed-in wrist pins, read ahead to *Piston/Rod Disassembly & Assembly*, page 71.

If a rod has excess pin-to-bore

Mounted on rod honing machine, indicator is preset to specification with micrometers. Connecting rod is then rotated on mandrel to indicate big-end ID. Each indicator division represents 0.0001 in. (0.002mm), so small discrepancies can be found.

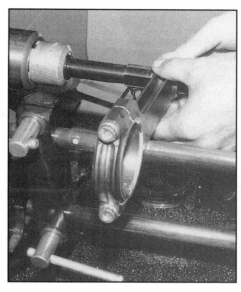

New pin bushing is honed to size. Pin-to-bore clearance for full-floating pins is in the 0.0001–0.0005-in. (0.002–0.012mm) range. Work toward the lower limit for a street engine; higher limit for high-performance and racing engines.

clearance, the small end of the rod will have to be rebushed and honed to achieve the desired clearance. To do this you'll first have to remove the bushing. You can do this with either a cope chisel or press, but it's best to leave this to the machine shop. The new bushing must be pressed in and honed to size anyway.

A special driver is used to press out the bushing. It pilots in the small end of the rod and has a narrow shoulder to butt against the bushing. While the bushing is pressed out, the small end of the rod must be supported squarely on its backside.

To remove the bushing with a homemade tool, cut a long 3/8-in. or larger bolt that is not fully threaded. Cut off the threads. Grind the end of the bolt so it's tapered and the point is crescent shaped.

Carefully work the point under the bushing and peel the bushing out. Be careful not to damage the rod. A gouge will create a stress riser and possibly cause problems when installing the new bushing. A split bushing will roll out easily at the seam. For a seamless bushing, work one side, collapsing the bushing inward by driving the chisel behind the bushing. Once the press on the bushing is relieved, push it out.

To install a new pin bushing, clean the small-end bore and check the bushing. One end of the bushing should be chamfered. The chamfer provides a lead-in when pressing in the bushing. If it isn't chamfered, make one with a knife, bearing scraper or file, page 113.

Use a press or vise to install the new bushing. If you are using a press, use the same tool you used for removing the old bushing. As for the vise, it should have soft jaws or jaw covers made of brass or aluminum. Press in the bushing so it's nearly centered in the small end.

If the rod and bushing have oil holes, line up the bushing oil hole with the one in the rod so they will match when the bushing is pressed in. If there is a wrist-pin oil hole in the rod, it's at the top of the small end. If the bushing doesn't have an oil hole but the rod does, press in the bushing first. Drill the oil hole in the bushing through the one in the rod. Use the same size drill as the hole in the rod. Protect the bushing from drill-point damage by inserting a piece of wood in the bushing prior to drilling.

After drilling, the bushing must be honed or reamed to match the wrist pin. Check that the oil hole isn't burred over from the sizing operation. If it isn't, clear the hole by turning the drill bit with your fingers and wash out the bushing with solvent. Once you've sized the bushing to the wrist pin, keep the piston-and-pin set with the rod. Using a bearing scraper, deburring tool or pocket knife, cut a small chamfer at both ends of the bushing ID. This will provide a lead-in for installing the wrist pin.

Check wrist pin fit by sliding it into the bushing. It should be snug, but slide through easily. If the fit is wrong, correct the problem before you move along. A loose pin will be noisy and a tight one will seize in the bushing.

Piston/Rod Disassembly & Assembly

If your engine uses pressed-in wrist pins, read this section. If it uses floating pins, move on to the next section.

A press with the needed tooling is required to remove pressed-in pins. If you have this equipment, you're in business. Do the job yourself and save the money. But if you don't have a press and the tools, farm out the work to your engine machinist to separate the rods from the pistons and reassemble them.

Warning: Do not hammer on a wrist pin or drive it out with an air impact tool! You may get them apart, but you'll ruin the pin, piston and rod. You must push out a pin with a press.

Remove Pin—To remove a wrist

ENGINE BALANCING

The performance and durability of your engine can be maximized if it is balanced. This is critical if it will be operated mostly at high rpm. The same is true if you've done extensive bottom-end work or parts interchanging.

Up, Down or Around?--Components that should be balanced are those that reciprocate or rotate. Reciprocating parts move up and down in the bores. Such parts include pistons, piston pins, pin retainers, rings and small ends of the connecting rods. Rotating engine parts include the crankshaft, connecting-rod bearings and big ends of the connecting rods. Exterior rotating parts also influence engine balance. These include the flywheel or flexplate, crankshaft damper and clutch.

Balancing Process—The first step in balancing an engine is finding an engine shop that has the equipment and know-how to use it. For your part, you must deliver all of the parts to the balancer after the machine work is done and the pistons have been fitted to their bores and marked accordingly. This includes reconditioning the connecting rods, installing the rod bolts and machining and matching the pistons to the bores. Note: Pistons or rods are not balanced as an assembly, so wait until after balancing to assemble the connecting rods and pistons.

The balancer will start by weighing the pistons. After finding the lightest piston, he'll reduce the weight of the others to its weight. This is typically done by removing material from the pin boss. Although the process is more involved, the same goes for connecting rods. The big and small ends of each rod are supported at their centers with the connecting-rod in a horizontal position. Each end is weighed until the lightest big end and small ends are found. The ends are then reduced to the weights of the lighter big and small ends by removing metal from the balance pads at each end. A balance pad is on top of the pin boss and the other under the bearing cap.

The small end of a connecting rod reciprocates with the pin and piston; the big end rotates with the crankshaft. The beam connecting the big and small ends contribute to both reciprocating and rotating weight in proportion to how close it is to the piston end or crankshaft end, respectively. Reciprocating weight is found by weighing the small end. This weight added to that of the piston, piston rings, wrist pin and pin retainers, if used, is total reciprocating weight. Rotating weight of the connecting rod is found by weighing the big end. This and the weight of the bearings, estimated weight of the oil between the bearings and crank combine to make total rotating weight.

Now for balancing: Bobweights, or canisters filled with lead shot, are clamped to the crankshaft rod journals to simulate rotating weight and a portion of the reciprocating weight as determined by a formula for that particular engine configuration. For example, balance of in-line and flat six-cylinder engines and 60° V12s do not require the use of bobweights, so the various parts can be balanced by themselves. Likewise, some engines are not affected by reciprocating weight, so only rotating weight is added to the canisters. V8s, on the other hand, require bobweighting. To balance the crankshaft, it is spun between centers similar to how a wheel-and-tire assembly is balanced.

Crankshaft counterweights counter, or balance, all of the rotating weight and, in some cases, reciprocating weight. When lighter connecting rods are used, as would be the case with aluminum, material is removed by either drilling or shaping the counterweights. If heavier rods are installed, weight is added to the counterweights. This is done by first drilling holes in the sides of the counterweights then pressing in short bars of heavy metal.

Internal & External Balance—Internally balanced engines are counterweighted by the crankshaft counterweights. External balancing is done by adding counterweights at each end of the crankshaft outside of the crankcase. This is usually required if there's not enough room in the crankcase for larger counterweights on the crankshaft or heavier connecting rods were installed at the factory on a crankshaft originally counterweighted for lighter connecting rods. In either case the damper and flywheel are counterweighted—made out of balance—to put the complete bottom end assembly in balance. Consequently, the crankshaft must be balanced with the bobweights, damper and flywheel.

To convert an externally balanced engine to an internally balanced one, the damper and flywheel are balanced by themselves so they will be in zero, or perfect, balance. This usually requires the addition of heavy metal to the crankshaft counterweights as described above. The advantage of making this conversion is to eliminate the bending forces caused by counterweights at the extreme ends of the crankshaft, thus reducing bending stresses on the crankshaft. Although not so critical on a low-rpm street engine, eliminating external balancing is very important on a high rpm performance engine. Another advantage of making the externally to internally balanced conversion is the damper and flywheel are easily balanced in the event either must be replaced.

Disassembling piston and connecting rod with pressed-in pin requires a hydraulic press and mandrels. Never drive out wrist pin with a punch and hammer.

pin correctly, support the piston on a hollow mandrel so it sits squarely on its pin boss. This is so the pin will have clearance to pass through as it's pressed out. For the same reason there must also be a clearance hole in the press bed.

A punch must be used between the pin and press ram. The best type of punch to use is one that pilots in the pin and shoulders against the pin. This is so the punch doesn't move off center and score the pin bore as the pin is pushed out.

To push out the pin, center the mandrel, pin and punch under the press ram. Bring the ram down and check that everything is square to the press bed and ram, especially that the mandrel is correctly positioned under the pin boss. Now push out the pin. When the pin breaks loose you'll hear it pop.

Install Pin—There are two ways to install a pressed pin: with force or with heat. Using force with a press is simple enough; the pin is pressed into a bore that's slightly smaller using basically the same procedure as removing the pin. By applying heat, the small-end bore is enlarged, allowing the pin to be installed by hand. When it cools, the press fit returns, allowing the small end to clamp tightly on the pin.

Although more critical, pressing in a pin is basically the reverse of pressing it out. You must be extra careful to support the piston and to line up the piston, pin and rod. If you don't, there's great risk of damaging the piston. In addition to supporting the piston firmly under its boss, apply some moly grease to the pin. This eases the installation and reduces the chance of galling the pin. Push the pin in far enough so it's centered in the rod.

Heating the small end of the rod is the method I prefer. Not only does it eliminate the chance of galling the pin and damaging the piston, it is quicker. However, because this is a very difficult job to do right, and because it is easy to ruin your rod before you know it, I recommend that you leave this job to the pros.

Orienting the Assembly—But let's suppose you are a pro, or feeling confident, and you are sure you can heat the rod and center the pin quickly without damaging it. If you insist on doing this job yourself, the first thing to do is arrange the connecting rods and pistons so it'll be easy to orient them correctly during assembly. This is very important: Piston-to-rod orientation must be correct, otherwise your engine will self destruct on or shortly after startup.

All pistons, except most high-performance pistons, have their pins offset to the right or toward the major thrust surface of the piston. Note: This is for engines using conventional, clockwise rotation. For those using opposite rotation, such as many engines built for marine applications, the pins are offset to the left. Pin offset is to load the thrust surface against the cylinder wall to keep the piston from rocking as it goes down the bore, thus prevent piston slapping and the resulting noise.

Many pistons with shaped domes must be correctly oriented, too. Put in backwards, the piston would collide with the combustion chamber and valves when it approaches TDC. Symmetrical pistons will clear the valves and combustion chamber regardless of their orientation, but you must still consider pin offset.

How do you determine which way a

Pistons, pins, connecting rods and new Spiralox retainers are organized prior to assembly. Pin retainers should never be reused. Install new ones. Note board for organizing rings for V6 racing engine.

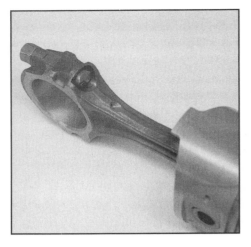

Some connecting rods have an oil hole in beam for lubricating thrust-side of cylinder wall; others a groove in bearing-cap parting line. Because of this and a chamfer on one side of big end needed to clear bearing-journal radius for rods sharing one journal, connecting rods must be oriented correctly to pistons.

piston must be installed? Simple. The manufacturer either casts a dot or notch in the piston top to indicate front or stamps an arrow with the word FRONT next to it. For pistons with shaped domes, orientation is sometimes indicated with a dot to indicate one side, not the front! Make sure you read the directions when installing new pistons. It's pretty difficult to goof up piston orientation in this case.

As for connecting-rod orientation, you'll have to be aware of more than one feature if yours is a V-type engine. For instance, connecting rods that use shared journals—two rods per crank throw rather than one per throw—have a large chamfer at one end of the bearing journal and a small one at the other end. This large chamfer must install against the crank cheek so it clears the journal radius. Otherwise the bearing will edge-ride the radius.

With V-type engines, rod numbers install in the direction of their cylinder banks: rods in the right bank install with numbers to the right and vice versa. Similarly, those with oil squirt holes are oriented with these holes aimed so they lubricate the opposite cylinder banks, just the reverse orientation of the numbers. A squirt hole may be drilled through the big end between the beam and bolt hole or it may be a notch in the cap. There may be no squirt hole at all. Oil thrown off the journals made possible by rod side clearance may be sufficient lubrication.

To prepare for assembling rods and pistons, arrange them in a pattern so you'll orient them correctly with minimum confusion. This is especially important if you are using heat rather than a press. You won't have much time to think about which way they go. Additionally, the piston will be upside down, so you'll have to think upside down. Confusing. Get everything sorted out in your mind before you begin assembling!

After you've assembled the first rod and piston, check it. Hold the assembly right side up and so the front of the piston points away from you—dot or arrow to the front and numbers to the right or left. Also check that the rod is centered in the piston and the pin is in its bore the same amount at both ends. If all is OK, great. Continue and make a final check once you've finished assembling the rods and pistons. Orient one wrong and you'll have to press out the pin and reassemble them correctly, increasing the chance of damage.

Note: Floating pins must be installed with new retainers, whether they are the wire, Truarc or Spiralox type. When installing, the convex side of Truarc or Spiralox retainers must go against ends of the pins. Also, it's best to have zero clearance between the end of the wrist pins and retainers.

TIMING SET

You should've already checked for timing chain and sprocket wear or gear tooth wear as described in Chapter 3, page 36. If you have, you should know whether or not you'll be replacing them. I suggest that even though your timing set checked out OK, replace it if you want your engine to be at least as good as it was when new. Assuming that you'll be replacing the timing set, what you need to know now is how to get the most for your money.

Chain & Sprocket

There are two basic types of chain-and-sprocket timing sets: the silent chain and the roller chain. There are sub-classifications of each.

Shape of piston dome requires that piston is oriented correctly in bore to prevent piston-to-valve and cylinder head contact.

Roller timing-chain set has three keyways in crankshaft sprocket. This allows cam to be timed as designed by cam manufacturer, advanced 4° for more lower-rpm torque or retarded 4° for higher-rpm operation.

The silent chain uses steel plates that pivot on pins. As the chain wraps around its sprocket it literally forms gear teeth. This type of setup is very strong and has been used for years. The difference here is in the cam sprocket.

With the silent chain an all cast-iron or a composite aluminum/nylon sprocket is used at the cam—a cast-iron or powder-metal sprocket is used at the crankshaft. The cast-iron cam sprocket is more durable, but the aluminum/nylon one is quieter. The problem is the nylon teeth tend to crack and break off over time and fall into the oil pan. Two undesirable things can happen as a result. The chain can jump teeth, knocking the cam out of time with the crankshaft. Either the engine won't run at all or it will run very poorly. Secondly, the nylon teeth will get drawn up against the oil pump strainer screen, blocking oil flow to the engine and causing a drop in oil pressure. I can leave the rest to your imagination.

Double Roller—A more durable setup is the double-row "roller" chain. It's similar to a bicycle chain, but with two rows. Although this timing set isn't as quiet, some OEM manufacturers have gone to it for its durability in high performance and heavy-duty applications.

If you opt for the "roller" chain, you have a choice: a chain with rollers that don't roll and one where they do. I suggest you go with the type that does. Although many engine component manufacturers offer the rolling roller chain, Cloyes True Roller chain-and-sprocket set is the best known. In addition to its true roller feature, the crank sprocket has two additional keyways, one for advancing the cam 4° for more bottom-end torque and retarding it 4° for more top-end power. If you want even more sophistication, Cloyes offers a True Roller setup where cam timing can be fine tuned around the three basic timings. Forget this last timing set, though, unless you will be dynoing your engine. Otherwise you'll be guessing and chances are you'll guess wrong.

Gear Drives—I'm jumping the gun on gear drive cams, but I don't want to lose those of you with engines using chain-and-sprocket drives. You may be able to modify your engine to accept a gear drive, although it'll be expensive. Many aftermarket suppliers offer them. The big difference here is an idler gear installed between the crank and cam gears. This is so that the cam will rotate in the same direction as it would with a chain drive.

As for the advantage in going to a gear drive, you'll eliminate the problem of chain "stretch"—which is elongation due to wear. This is an awful expensive approach, though. The reason most go to gear drive cams is for the noise! Street rodders like the blower-drive-like whine coming from underneath the hood of their '32 Ford, '55 Chevy or whatever.

If your engine was equipped with a gear-driven cam, the cam gear may have been a phenolic resin/steel-composite running against a steel gear at the crank. This setup makes for quiet operation, but you can go to the more durable cast-iron cam gear with only a slight increase in engine noise, but improved durability.

OIL PUMP

Without exception, the oil pump is one of the most durable components in an engine. The reason is obvious. If the oil pump fails, engine failure is not far behind. This in itself doesn't make an oil pump durable, but it does

Checking oil-pump clearances is done after it passes visual inspection—no scoring or grooving on rotors or housing. Internal-to-external rotor and internal rotor-to-housing clearances are checked on this G-rotor pump. Before removing them, note how rotors are oriented. They must be reinstalled in their positions to the housing and each other.

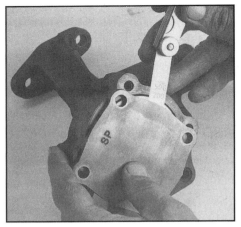

Cover plate is held in position over rotors while checking rotor end clearance.

Make sure pressure-regulator valve moves freely. If it feels sticky but pump passed all other inspections, free valve by soaking pump in lacquer thinner or carburetor cleaner and working valve back and forth.

make oil pump durability a major consideration during engine design. And it doesn't hurt that the oil pump is drenched in oil during engine operation. Consider, though, that this oil is basically unfiltered except for big chucks that are removed by the pickup screen. If these particles are larger than internal clearances in the pump, wear and scoring result.—or worse, pump lockup.

All things considered, don't think that oil-pump inspection isn't necessary. For example, engine rebuilders replace the oil pump with a new one as standard practice. Just the same, oil pump problems are relatively few, particularly with a well-maintained engine—one that's been treated to frequent oil and filter changes.

High Volume Pumps

Some advice: If the original oil pump checks out OK, use it; if not, replace it with a stock oil pump, not a high volume pump. The reason is simple: A high volume pump wastes horsepower pumping oil that the oiling system doesn't need. If you were wondering, the main difference between a standard pump and a high volume one is the gears or rotors in the later version are longer or have a larger diameter and more teeth. This pump then displace more oil for each revolution. Once regulated pressure is reached, surplus oil is routed directly back to the oil pan without going through the lubrication system, thus wasting power to pump it. So why are high volume oil pumps available?

High volume pumps are used only to maintain oil pressure in engines that have increased oil clearances due to high wear or are built for racing to provide additional cooling. In these cases, wider clearances increase the need for additional oil to be pumped to maintain the desired pressure. Consequently, the volume of oil being whipped around in the crankcase is increased. However, being in the horsepower business, many high performance engine builders have gotten wise and don't build in excessively high clearances and, consequently, don't need power-absorbing high volume pumps. This reduces wasted horsepower, load on the pump drive and oil foaming.

As for restoring and inspecting the original pump, start by scrubbing the outside of the assembly with a stiff bristle brush in solvent. This will keep debris out of the pump when you disassemble it.

Wash out the pump internals by submerging the pickup in clean solvent and rotating the pump with its driveshaft until solvent gushes from the pump outlet. If rotating the pump clockwise several times doesn't cause it to pump solvent, simply turn it counterclockwise. Once solvent flows, it won't take long for it to become clear. The pump is now clean inside and you can stop turning.

Teardown & Inspect Pump—With the pump over a drain pan, hold it in several positions to allow solvent to drain. If the pump has a bolted on pickup, remove it to make handling

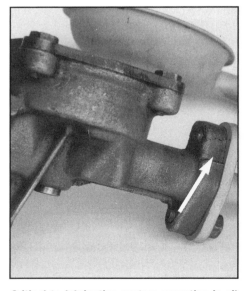

Critical to lubrication system operation is oil pump pickup. Clean screen, then attach pickup to pump. This one bolts on. Crack through flange at bolt hole (arrow) in new pump didn't appear until pickup was bolted on. Disaster could have resulted if crack wasn't discovered, so never stop looking.

easier—don't go to the trouble of removing a pressed-in or screwed-in pickup.

You'll have either a gear pump or rotor pump. The first has two external gears—a driving gear and a driven gear. The rotor pump has one internal rotor and one external rotor.

With the pump positioned over a bench, remove the cover screws or bolts. Be careful that the gears or rotors don't fall out onto the floor and get damaged. Before you remove the gears, mark them with a felt-tip marker so they can be returned to their same positions. All you'll need to know about the rotors is which end of the driven, or outer, rotor installs into the pump. Even though both rotors should be marked in some manner, mark them anyway just to make sure.

Inspect the pump body and gears/rotors. Deep scoring, pitting and chipped teeth indicate the pump ingested large metal or dirt particles—replace the pump. If you don't find anything but minor scratches, it's OK to reuse the pump providing clearances check out OK.

End Clearance—Regardless of the type of pump gears or rotors used, gear or rotor end clearance should be in the 0.001–0.004-in. range. To measure this, the gears or rotors must be in place. Now, hold the cover firmly over the pump body so it covers about half of the gears. Using feeler gauges, find the one that fits snugly between each gear and the cover. If it's in the 0.001–0.004-in. range, end clearance is OK. Consider 0.0025 in. clearance ideal if the pump body is cast iron; 0.0015 in. if it's aluminum. Why the difference? Because the thermal expansion of aluminum is greater, end clearances with both pumps should be the same at their operating temperatures.

You can also use Plastigage or a depth gauge for checking end clearance. With Plastigage, just lay a strip on the end of each gear, install the cover plate and remove. Check end clearance using the Plastigage sleeve just as for checking bearing clearances, page 119. However, if you have a depth gauge, use it. It will yield the most accurate results.

MODIFYING YOUR OIL PUMP

End Clearance—If everything checks out OK, but pump end clearance is excessive, you can easily correct it by sanding the pump body. You'll need a sheet of 220 wet-or-dry sandpaper and a section of thick glass to provide a rigid flat surface on which to sand.

With the sandpaper on the glass and wetted with solvent, position the pump body mounting face against the paper. Move the pump body back and forth, trying to remove the same amount across its face. Do this by holding the pump body square to the glass surface and moving it back and forth as you press evenly and lightly—don't let it rock. Do this by positioning your fingers as close to the sandpaper and low on the pump as possible.

Check your progress with feeler gauges and compensate as you sand so you don't remove too much metal and that it is removed evenly across the pump face. Also, keep the sandpaper washed off and rewet it with clean solvent.

Before you install the gears or rotors to check end clearance, wash off the pump body to remove the metal and abrasive grit. These particles will affect your measurements if not removed.

If end clearance is under the optimum—0.0025 in. for cast-iron bodies and 0.0015 in. with aluminum—increase clearance by performing the same sanding operation on the gears or rotors. Remove material from the ends the cover installs against. Here you can use outside mikes to check your progress. Just subtract the amount you wish to remove from the thickness of the gears or rotors and work to that dimension. Don't forget to clean them before miking their thickness.

One easy and beneficial thing to do is to break, or chamfer, all sharp edges of the pump body and gears/rotors. Use a very small, fine-tooth file to do this. Afterwards, flood the pump and gears/rotors with solvent to remove the filings.

Secure Pickup—If the pickup falls from the pump you can be sure engine destruction will quickly follow. To prevent this from happening make sure the pickup is positively secured to the pump body. If the pickup is pressed into the pump, secure it by brazing. Before you do this, remove the pressure relief valve and spring. If The pickup is bolted to the pump, cross-drill the bolt heads and safety wire them. As for pickups that thread into the pump, no additional steps are required to secure them. Just make sure the pickup will be square to the bottom of the oil pan when installed.

Pressure-Relief Valve—About all you have to do here is make sure the valve moves freely. Use a small screwdriver or scribe to push on the valve, which may be a cylinder or ball. If it moves without binding, it's OK. If it doesn't, soak in lacquer thinner or carburetor cleaner to loosen. ■

CYLINDER HEADS 5

Of all major assemblies, the condition of a cylinder head usually has more to do with engine performance than any other. And wouldn't you know it? Except maybe for the valvetrain, a cylinder head is more likely to need attention before components such as the pistons, crankshaft or connecting rods. So it's not uncommon that a top overhaul alone will correct an engine's performance problems. But beware: Increased combustion chamber pressures due to better valve sealing usually cause excessive blowby—leakage down past the pistons. This is more likely to occur with high-mileage engines.

I've included cylinder-head teardown, inspection, reconditioning and assembly in one chapter. I've done this because that's the way cylinder-head reconditioning is usually done: Once cylinder head teardown begins, an engine shop typically carries cylinder head work through to completion. The assembled heads are then set aside while the rest of the engine is reconditioned.

Do It Yourself or Farm It Out?—Although I cover cylinder-head reconditioning in detail, it's best that you farm out this work. Why? The expensive, specialized equipment is the first reason. At the minimum

Reconditioned head was restored to better-than-new condition. Work included resurfacing, new valve springs, reconditioned valve guides, three-angle valve job, minor port work and reground valves.

you'll need a valve grinder, valve seat grinder or cutters and guide-reconditioning equipment to do the job right. You can't rent such equipment and, unless you have won the lottery, there certainly isn't any justification to buy this expensive equipment for doing one rebuild—or even an occasional rebuild.

In the event you have access to a shop full of equipment, you must know how to use it. It's a lot easier to tell someone to recondition valve guides than to do it yourself. So, if you don't have the tools or equipment, take your cylinder heads to a good machine shop and have them do the job you want. It won't take long nor will the job cost that much if you consider the time you've saved. If you opt to go this route, you should still read this chapter. What I tell you here will give you a working knowledge of the process so you can better choose a cylinder head rebuilder and explain exactly the work you want done.

Rocker arms use common shaft for mounting. When removing rocker arm assembly, keep all parts organized so they can be returned to their original positions. This includes stands and bolts.

TEARDOWN

Start out your cylinder head work by soaking the heads with degreaser. Remove the loosened deposits with spray from a garden hose or avoid making a mess at home by taking the heads to a local car wash. Once there, apply the degreaser, then use the blast from the high pressure spray to remove baked-on deposits. A more thorough cleaning will come later.

Begin by removing the spark plugs now. Don't discard them, though. They will come in handy for sealing your engine at assembly time. Also, they'll work great for "masking" the plug holes when painting your new engine.

Rocker Arm & Pivot Relationship

If your engine uses individually mounted rocker arms, they should still be on the heads. It's now time to remove them, but heed this warning: Keep the rocker arms and their pivots together. They are matched sets. Mix them up and they will gall, sending metal particles to the oil pan and through the oil pump.

To keep the rockers and pivots together, slide them over a 2-foot length of wire. Loop the wire through the first rocker and pivot, then slide on the others; rocker and pivot, rocker and pivot . . . until they are all on. Additionally, if the valves don't need to be reconditioned, you must keep the rocker arms in order so they can be reinstalled on their valves.

Valve Removal

Keep the valves and springs in order as you remove them. If the valve guides don't need reconditioning, this is a must. If they do, keeping the valves and springs in order doesn't matter. The problem is you may not know whether or not they do. Like other moving parts in an engine, valves and guides wear in together, so they should be kept together unless one or the other is replaced or remachined.

To do this, either drill the same number of holes as there are valves in a length of wood—a yard stick does nicely—or punch the holes in the bottom of a cardboard box. Make two rows for two heads and label them. Mark the holes so you'll know which head and guide each valve came from.

With two heads, mark each one **R** and **L** with a letter or center punch and indicate likewise on your valve holding "fixtures." Mark the heads with letter punches or a center punch. Such marks won't be removed by cleaning in engine cleaner as paint would.

As I said earlier, this extra care is unnecessary if the guides are reworked or the valves are replaced. But saving the valves if at all possible is a good idea because they are expensive.

Remove Valve Springs

You'll need a valve spring compressor to remove the valves. The most common is the C-type, so called because of the shape of its frame. A fork-type compressor can also be used, however it is for use with a head installed on the engine.

A C-type compressor straddles the head. The C-clamp like screw end butts against the valve head and the quick-acting end fits over and around the spring retainer. As the compressor lever or screw is operated, it forces the retainer down, compressing the spring against its spring seat. This exposes the keepers—collars, keys or locks, if you prefer—allowing them to be removed.

These half-round, wedged shaped, devices with one or two raised beads on their ID wedge between the retainer and valve tip, locking the retainer to the valve as the spring pushes up. The bead(s) fit in a groove(s) in the valve tip and the wedge fits into a tapered bore in the center of the retainer.

Keepers seem to "grow" to their retainers over time and with use. Consequently you should break the retainer loose from the keepers before compressing a spring. If you don't, the compressor may be damaged if you force it. Use the following method for

With a wood block under valve head, break loose spring retainer by striking socket placed against retainer with soft-faced hammer. Be careful. This may release keepers, allowing retainer and spring to fly up.

After breaking loose retainers, compress valve spring just enough to allow removal of keepers. You'll need a valve spring compressor to do this.

doing this; as you begin to compress a spring or before you attempt to compress it. Just as you begin to apply pressure to the spring with the compressor, tap on the retainer/compressor with a soft mallet. This will break it loose from the keepers, allowing you to compress the spring and release the keepers.

The next option for breaking loose valve spring retainers is used by most engine rebuilders. The reason is simple—speed. It's a lot faster and, if done correctly, a spring compressor may not be needed to release the keepers and retainers from their valves. Sound appealing? Here's how to do it.

Find a small block of wood, one that fits into the combustion chamber and butts squarely against both valve heads. Position the head with its gasket surface down. Fit the block of wood under the valves of one combustion chamber. Now, using a 5/8- or 3/4-in. deep socket, set it on the retainer and check its fit. It should butt snugly against the retainer. You'll also need a brass, lead or plastic mallet.

The problem here is retainers tend to go flying when released in this manner, so proceed with caution. Start by laying a rag over the spring and retainer to catch any wayward keepers. Position the socket against the retainer and strike it with the mallet. Keep a firm grip on the socket just in case the keepers completely release.

Remove all retainers and keepers, but don't concern yourself about keeping them in order. There is a minor exception, though. Some exhaust valves use a special retainer called a rotator. Its job is to rotate the exhaust valve so it seats in a different position on its seat every time it closes. This improves cooling and, thus, increased exhaust-valve durability.

Remove Valves

Once the retainers, springs and related hardware are removed, slide the valves from their guides. This may not be so easy on high-mileage engines as the tip of the valves may be burred over. Fix this by placing a flat file at an angle against the valve tip. Now rotate the valve while stroking the file against the edge of the tip. This will remove the burr, allowing the valve to slide out of its guide. Also, if the valves use O-ring seals installed in a groove just below the keeper groove, remove them with needle-nose pliers or a scribe.

Some valve spring assemblies consist of an inner and outer spring or a spring and damper. The damper is a flat wire wound to fit tightly inside the larger spring. The damper and small spring accomplish similar functions: Each controls the valve in a manner similar to how a shock absorber controls the suspension of a car. The difference is the inner spring also adds stiffness to that supplied by the outer spring. A few high performance engines use two springs—an inner and outer—and a damper.

Store the springs and, if used, shims, dampers and spring seats so you can reinstall them in their original positions. If there's no seat and valve work done, they should go back in their original positions. But if the seats or valves are reconditioned or replaced, pairing is not necessary.

Spring Shims/Seats/Rotators

Some valve assemblies use one or

more shims between the spring(s) and head. These thin washer looking discs are installed under the springs to restore height and/or spring "pressure." You may also find a spring seat that installs between each spring and the head, particularly if the head is aluminum. In this case they are a must. These hardened steel discs prevent the springs from "eating" through the soft aluminum spring pads.

One other component sometimes installed under exhaust valve springs is a rotator. This accomplishes the same thing as the combination retainer/rotator, but it moves the rotator assembly from the inertia of the valvetrain. In plain English, the cam lobe and valve spring don't have to move it during valve opening and closing, thus reducing the load on the valvetrain. It also increases the usable rpm of the engine because valve float will occur at a higher rpm assuming all other factors are equal.

CLEANUP & INSPECTION

Before inspecting or reconditioning a head, thoroughly clean it. The best

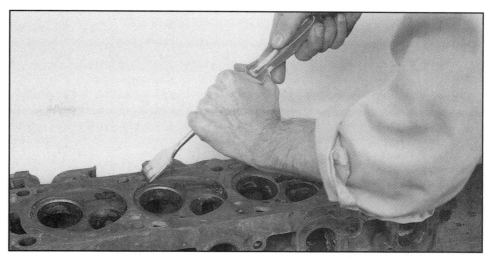
To ensure sealing, head gasket and other gasket surfaces must be clean down to bare metal. Be careful when scraping aluminum cylinder heads. Use chemical gasket remover and plastic scraper on aluminum. A large flat file is used to smooth over imperfections and remove small bits of gasket material.

and easiest way to do this is at the machine shop. If you haven't shipped the block and its parts to the shop, deliver the heads to have them cleaned at the same time. Otherwise, you'll have to degrease and decarbonize them yourself.

Once you have all the grease and grime off, remove carbon from the combustion chambers and ports and gasket material from all sealing surfaces. An old screwdriver or small chisel is handy for removing heavy carbon deposits. Just be careful not to damage the valve seats, spark plug threads or head gasket surface.

Caution: For aluminum heads, be very careful, particularly with the head gasket surface. Aluminum is easily damaged.

Finish the combustion chamber cleaning job with a rotary wire brush chucked in a 3/8-in. drill motor. This will put a light polish on the surface so you can spot cracks easier.

Turn your attention to the gasket surfaces. Here, a gasket scraper works best. A putty knife will also work, but unlike a gasket scraper it will flex when pressure is applied to it. Whatever you use, be very careful not to gouge the head gasket surface. This could provide a path for combustion-chamber gases to escape.

When you think you have the head gasket surface "perfectly" clean, run your hand over it. You'll be surprised by how much you didn't get. All gasket material must be removed. When you've reached that "perfect" stage, run a large flat file across the head. High spots will show up as bright areas. This will also remove any light burrs, remaining gasket material and other small

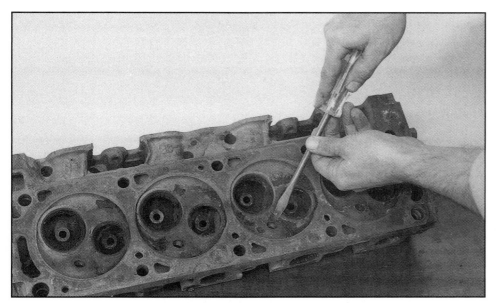
Rotary wire brush in drill motor works best, but blunt screwdriver and hand-operated wire brush will get carbon out. Again, be careful when working with aluminum. Just be careful.

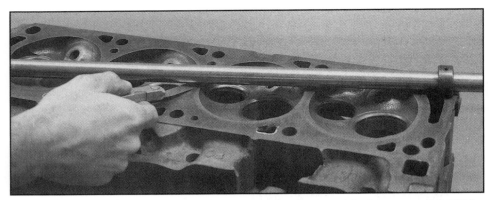

Check head surface flatness with a precision straight edge and feeler gauge. If 0.004-in. (0.10mm) feeler gauge fits under bar, head should be resurfaced. For an inline six and eight, make it 0.006 in. (0.15mm) and 0.008 in. (0.20mm), respectively.

imperfections.

Don't forget the water passages. Use a pocketknife or small screwdriver to remove any large deposits, then finish up with a round file. To remove all that junk that fell into the head, turn it upright over some old newspapers or a trash can and move it around until the loose debris falls out.

Head Warpage

For a cylinder head to seal best against the block, the gasket surfaces must be straight. But if the heads are cast iron and you're sure the engine has not been overheated, chances are they aren't warped. Check them anyway. Repeated heating and cooling over the years from startup heating and shutdown cooling can warp the best cylinder head. To check a head for warpage you'll need feeler gauges and a precision straightedge.

Make sure the cylinder-head gasket surfaces and straightedge are perfectly clean before checking warpage. If there is any gasket material or dirt left on the head you won't get an accurate reading. So run your hand over the head-gasket surface once more. Also, run your hand across the machined edge of the straightedge. Wipe both surfaces thoroughly immediately before making your check.

Generally speaking, a head that's warped more than 0.004 in./0.10mm (0.006 in./0.15mm for an inline six or 0.008 in./0.20mm for an eight) should be resurfaced. With this in mind, set the straightedge lengthwise on the head. Try to slip the feeler gauge between the straightedge and head. To make this easier, position a light on one side of the straightedge and look at it from the opposite side. If you see light between the head and straightedge, try slipping the feeler gauge under the straightedge at that point. If a 0.004-in./0.10mm feeler gauge slides underneath, try a thicker gauge. For example, if a 0.008-in./0.20mm feeler gauge fits, then the head is warped 0.008 in./0.20mm. It should be resurfaced.

Check the head lengthwise from side to side across the head, paying particular attention to the narrow section between each combustion chamber. Place the straightedge diagonally across both corners to check for twist.

What's considered too much warpage depends on the type of head gasket being used. For instance, the thicker composition-type gasket used with most engines as opposed to metal shim-type gaskets will tolerate some irregularity between the block and head. Modern composition gaskets use materials that replace asbestos such as Kevlar or fiberglass. They are further reinforced with steel, copper or aluminum. So if this is the type of gasket you'll be using, an additional 0.002 in./0.05mm of warpage can be tolerated.

A shim gasket is another matter. It's made of a sheet of steel, copper or aluminum, meaning it can't compensate for much if there's any warpage. Limit total warpage to 0.002 in./0.05mm with a shim gasket.

Cylinder Head Milling

If one or both heads of a V-type engine are warped, both must be milled the same amount. This is so both banks will have the same

Resurfacing cylinder head. If more than 0.020 in. is removed from head that goes on a V-type engine, bottom and sides of intake manifold should be resurfaced to maintain intake-port alignment. There are other considerations: Compression may be raised to too-high a level and shorter pushrods may be needed if rocker arms are not adjustable. Valve-to-piston clearance will also be reduced.

Setting up intake manifold in preparation for resurfacing. Only one manifold-to-head surface is milled. For manifold that seals to top of block, bottom side should also be resurfaced. How much material that should be removed from side and bottom of manifold directly relates to that removed from cylinder heads.

Although combustion chamber shape and stroke-to-bore ratio affect changes in compression ratio by milling, there's approximately 1/10 of a ratio increase for every 0.030 in. removed from a head that has the average combustion chamber. So if an engine had a 10.5:1 compression ratio and 0.060 in. was removed from the heads, an 11:1 compression ratio will result. Don't forget, too, that the valves will get that much closer to the pistons. For more on compression-ratio, turn to the sidebar on page 103.

Rocker Arm Stud Replacement

A damaged rocker arm stud should

compression ratio. Mill just one head and engine idle will be rough due to unequal compression. Additionally, if more than 0.030 in./0.76mm is removed from the heads, the intake manifold runners and bolt holes will not align with those in the heads.

If more than 0.030 in./0.76mm is removed from the heads, a proportional amount must come off the sides. Additionally, if the manifold seals to the block, material must be removed from the bottom of the manifold. Refer to the chart for how much to remove from the sides and bottom of an intake manifold used with a V-type engine.

Milling Tips—Don't remove any more than what is needed to correct cylinder-head warpage. Take too much off, say 0.015 in./0.38mm, and some things may be negatively affected, such as: compression ratio, valve-to-piston clearance and pushrod length.

Many engines built prior to 1973 had very high compression ratios—average was 10:1, but 12.5:1 for high performance engines was not uncommon. Although good for power and economy, the drop in gasoline octane that followed made it impossible to find fuel these engines could run on. Avoid increasing the compression ratio of one of these engines.

INTAKE MANIFOLD VS. CYLINDER HEAD MILLING REQUIREMENTS (90° AND 60° BLOCKS)

Removed From Cylinder Heads	Removed From Manifold Bottom	*Removed From Manifold Sides
90° Block		
0.020	0.028	0.020
0.025	0.035	0.025
0.030	0.042	0.030
0.035	0.049	0.035
0.040	0.056	0.040
0.045	0.063	0.045
0.050	0.070	0.050
0.055	0.078	0.055
0.060	0.084	0.060
60° Block		
0.020	0.023	0.011
0.025	0.029	0.014
0.030	0.034	0.017
0.035	0.040	0.020
0.040	0.046	0.023
0.045	0.052	0.026
0.050	0.058	0.029
0.055	0.064	0.032
0.060	0.069	0.035

*Double amount when milling one side of manifold

Intake manifold milling figures are for those used with cylinder heads whose intake sides are square to head gasket surface. For those having different intake manifold mounting surface angles, consult your engine machinist to determine machining requirements.

Special tools are available for removing damaged rocker arm studs, but you can improvise by shimming with washers or rocker arm pivots stacked over stud under free-running nut. Liberally oil nut and stud threads to minimize friction. Tighten nut to pull out stud. Add to shim stack as stud pulls out.

be replaced. Replace one if the threads are damaged, it was notched by the rocker arm, it is broken or stud is loose in the head. The threads may be either stripped or worn so much that the adjusting nut doesn't have sufficient breakaway torque to hold rocker-arm adjustment. Notching occurs when the rocker arm tries to roll over on its side and contacts the stud because of another problem in the valvetrain. A notch is gradually worn in the side of the stud as the rocker arm works back and forth against it. The stud may break off if this continues unless it's found and corrected or some other valvetrain failure occurs first. A loose pressed-in stud would have partially or fully pulled out of the head.

Rocker Arm Stud Removal— How you remove a rocker arm stud depends on the damage and type of stud. For removing a pressed-in stud, a remover can be used if the threads are damaged or are even there. It installs over the stud and seats on the stud boss. As a nut is run down on the stud, it bottoms against the puller and gradually pulls the stud from the head. A ball bearing in the stud remover reduces effort required to turn the nut.

You can pull a stud without the use of a special remover. Simply stack up some washers on a socket set over the stud. Oil them to minimize friction, then run a nut down against the washers and socket. As the stud pulls out of the head, install additional washers under the nut to keep from running out of threads.

If the threads are damaged or the stud is broken off, the removal becomes more difficult. If the stud breaks off, the break normally occurs at the bottom of the threads. Consequently, you'll have to cut threads on the unthreaded portion of the stud with the appropriate die. Put as many threads on the stud as possible. Now, remove the stud using the same technique described above. But what if the stud breaks off short? If there isn't enough length for threading, the stud will have to be removed by drilling out. I recommend that you don't attempt removing the stud yourself. Instead, take it to an engine machine shop so you don't risk damaging the head. They'll set up the head on a valve seat and guide machine, find the center of the stud, and drill it out. If you attempt this, chances you will drill into the side of the stud bore.

For a stud that's stripped, it might have a portion of good threads that can be used for pulling. This being the case, you won't have many threads to work with so you'll have to pull the stud a little at a time in between adding washers.

Removing a threaded in stud is relatively easy. All you have to do is unscrew it unless it's broken off flush with the head. It will then have to be drilled out similar to how it is done with the pressed in type, so the same precaution applies—take it to an engine machine shop. To make the repair, they'll drill the stud, then remove it using a bolt extractor.

Rocker Arm Stud Installation— If the rocker-arm-stud bore for a pressed-in stud is in good condition, a standard diameter stud can be installed. Again, there is a special tool for driving in studs, but you can improvise. Just thread two or three free-running nuts on the stud so the last nut stands slightly taller than the stud. Tighten, or jam, the nuts against one another. Using a brass or lead hammer, drive in the stud until it is the same height as the other studs. This is particularly important with positive-stop rocker arms. To make this job easier, measure down from the bottom of the threads of an installed stud to the stud boss and transfer this measurement to the new stud with a scribe mark. Bluing the stud before makes the scribe mark stand out. Start the stud by lining it up with the hole and tap it in a little at a time. Stop driving in the stud when the scribe line is even with the top of the stud boss.

Oversize Studs—Stud installation is more complicated when you have to replace one that was loose. This is because the hole in the head will probably be enlarged. If you attempt to install the same size stud, it will also be loose and pull out. To correct the problem, an oversize stud must be installed providing you elect to stay with a pressed-in stud as opposed to the screw-in type. If so, check with your dealership or auto parts store to determine available oversizes and go with the smallest one—providing it's larger than the one that loosened. Mike the old one to make sure. Examples are 0.006-, 0.010- and 0.015-in. oversize.

Prior to installing an oversize stud, ream the hole to an interference fit of 0.0007-in., or so the hole will be 0.0007-in. smaller than the stud. This will require a special reamer, or a step reamer made just for this purpose. The reamer will cost considerably more than the stud, so consider farming out this repair. If you elect to do it yourself though, follow the same procedure for installing the oversize stud as was used for the standard stud.

Screw-In Studs—Now, for the method I prefer—even for replacing a pressed-in stud that doesn't require reaming. Simply buy a threaded replacement stud and tap the hole with compatible threads. When doing this, be careful. The tap must go in straight, otherwise the stud will lean and cause the rocker arm to be out of position. Once threaded, install two nuts on the stud, jam them together, apply some sealing Loctite to the threads and run the stud in so it aligns with the other studs. Sealant is required to ensure that coolant won't leak from the water jacket into the rocker arm area.

Converting to Screw-In Studs—For use with a high-performance cam, your heads should have screw-in studs. If it's not so equipped, consider making the conversion. This will require the services of a machine shop because the stud bosses must be milled lower to a precise height and

If original stud pulled out during engine operation or damaged one pulled out easy, ream for oversize stud. Ream should be 0.002-in. smaller than stud diameter. Thread on two nuts and jam them together. Run on top nut so it's flush with top of stud. Drive in stud with brass or lead mallet.

the stud holes tapped so alignment is maintained. Don't forget to seal the studs.

VALVE GUIDES & STEMS

Although cylinder head resurfacing is normally done at an engine machine shop, valve guide inspection and machining signals the beginning of true cylinder head reconditioning work. If such work is not done, a cylinder head is not reconditioned.

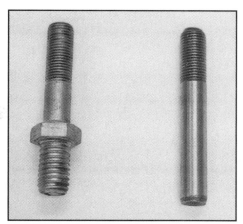
High performance screw-in rocker-arm stud compared to pressed-in stud. To convert from pressed-in to screw-in studs, stud bosses must be milled and stud holes tapped.

Cylinder head work includes valve, valve guide and valve seat reconditioning. And as I've already mentioned, such work requires special equipment and skills. If done incorrectly, it will ruin what would otherwise have been a successful rebuild.

If a valve guide is badly worn, it can't guide the valve squarely onto its seat. The valve will then wiggle and bounce from side to side before settling on the seat and closing. It may never close at higher rpm, resulting in combustion chamber pressure losses and the obvious drop in power output. If this condition isn't corrected, the

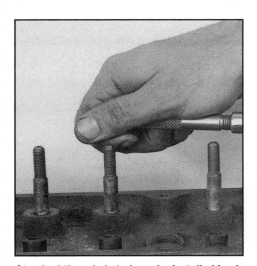

Standard threaded stud can be installed in place of pressed-in stud. Using a special tap, thread hole in stud boss. Tap has pilot so it follows hole during tapping to maintain stud alignment. Apply locking sealer to stud threads and run stud in with two nuts jammed together. Measure up from stud boss to place stud at same height as original studs.

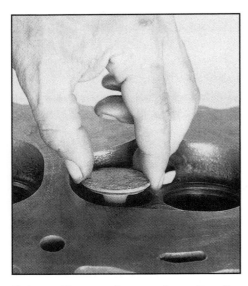

Not exactly accurate, experienced engine builder can judge if guide needs reconditioning simply by feel of how much valve wiggles in guide.

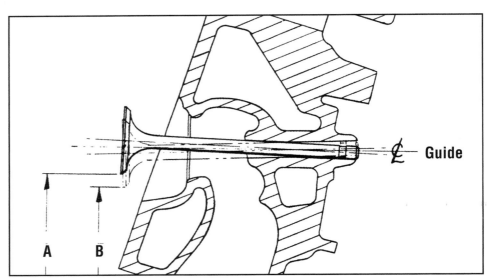

More accurate way of checking guide is to measure wear how much valve wiggles—difference between A and B measured at valve head. Divide this figure by 3.5 and you'll have approximate stem-to-guide clearance. Measure side movement too.

seat eventually beats out and guide wear continues at an accelerated rate. Additionally, a worn guide allows excess oil to pass down the guide between the valve stem and guide. This causes that blue smoke out the tailpipe, oil consumption and coke deposits on the spark plugs, back sides of the valves and in the ports. Such deposits can cause engine misfire and reduce the flow efficiency of the ports, resulting in additional power loss.

To head off such problems, let's look at reconditioning the valve guides, valve seats and the valves themselves.

Valve Guides

There's more than one way to correct valve-guide wear. Be aware that each machine shop has its preference. Machine shop A will bore out the old guides and press in new ones; machine shop B will use oversize valve stems; and machine shop C will only install guide inserts. It depends on where you live, a shop's equipment and the machinist's preference. I'll talk about each and you can make your choice. But first you must determine whether or not the guides should be reconditioned.

For street-driven engines the advantages and disadvantages of most reconditioning methods are academic. So don't waste your time searching for the hot tip in valve-guide technology. Instead, read about the different approaches available, choose the one you prefer after reading this chapter and find the shop that does it according to your preference. Don't try to persuade a machinist to use a method he is not comfortable with.

Valve-Guide Wear—Before you can decide whether or not the guides need reconditioning, you must determine how much they are worn. There are four ways to measure valve guide wear: with a dial indicator, wiggling the valve in its guide, with a taper pilot, and with a small-hole gauge and micrometer.

To check guide wear or clearance with a dial indicator, install a valve in the guide. Mount the dial indicator with its tip 90° to the valve stem or valve head in the direction you want to measure wear. Maximum wear usually occurs in the plane of the valve guide and rocker arm pivot, but maximum wear can be 90° to that direction if rail-type rocker arms are used. To be sure, check the guide in both directions and at both ends.

If you're measuring at the tip end of the stem, lift the valve about 1/8 in. off it seat and position the indicator tip—it should be a flat one—against the stem at 90° to it and close to the top of the guide. Push the stem away from the indicator and zero its dial. Push the tip end of the valve stem toward the indicator and read stem-to-guide clearance directly.

To check stem-to-guide clearance at the valve head end, raise the valve to its full-open position. You can tell this by the wear at the top of the stem. Positioning the valve in its guide is made easier by slipping a short rubber hose over the tip end of the valve so it aligns with the maximum wear mark. Or simply wrap some masking tape around the tip end and align it in the same manner. Raise the valve to its full-open position, set up the dial indicator with its tip against the margin of the valve, push the valve away from the indicator and zero it, then pull the valve toward the indicator. Stem-to-guide clearance will be approximately half of the reading.

Wiggling a valve in its guide to determine guide wear requires no equipment, just a lot of experience. This is normally how shops determine if guides need reconditioning—by feel. However, if you have a vernier caliper you can do the same thing, but more accurately. Start by positioning the head as shown in the nearby sketch—so the guides are level. With the valve pulled down out so its tip end is flush with the top of the guide, measure valve wiggle with the depth gauge end of your vernier caliper. Divide the amount of wiggle by 3.5 and you have approximate stem-to-guide clearance that would exist with a new valve.

Using a taper pilot is the least accurate way of determining guide wear. A taper pilot is a tapered pin inserted into the end of a guide until it is snug. The diameter where the pilot stops is miked to determine guide diameter at the top or bottom of the guide. The measured diameter less the specified guide diameter is considered guide wear—but it's not accurate. The pilot stops at the minimum distance across the guide rather than the maximum distance. It's the maximum distance you should look for, so don't use this method.

The most accurate way to determine guide wear is to measure with a small-hole gauge and a 0–1-inch/0–25mm micrometer. You'll need the C-gauge out of a set of four small hole gauges. Unfortunately, you'll have to buy the whole set to get one gauge. Fortunately they aren't expensive. Here's how to use a small-hole gauge.

Insert the ball end of the gauge in the guide and expand it until it fits with a light drag. Check the guide bore at several places up and down and around the bore to determine maximum wear. After setting the gauge, withdraw it and mike the ball

 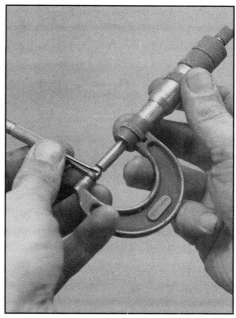

Direct measurement is the best way of determining valve guide wear. Small-hole gauge is first expanded to fit largest section of guide, then it is miked. If valve stem diameter subtracted from this figure is more than 0.002 in. (0.05mm), guides should be reconditioned.

end across its widest point. Subtract minimum valve-stem diameter from this figure and you have maximum stem-to-guide clearance. The point of maximum guide wear will be at the top or bottom, so measure around the guide at these points to find maximum wear, or diameter. Also measure for minimum guide diameter. If the difference between maximum and minimum diameters, or out-of-round, exceeds 0.002 in./0.05mm, the guides should be reconditioned.

Measure Valve Stems—There are three common valve stem diameters that fit 5/16-, 11/32- and 3/8-inch guides. Actual valve stem diameter is 0.001–0.004 in./0.03–0.10mm smaller than the guides. This allows for the needed stem-to-guide clearance. Exhaust valve clearance may be more than that for the intakes.

To get exact valve stem-to-guide clearance, you'll have to measure valve stem diameter at the point of maximum wear. Once you have this figure, subtract it from maximum valve guide diameter. You'll need a 0–1-in./0.25mm micrometer to measure the valve stem.

Maximum valve stem wear occurs at the top or bottom of the wear pattern on the stem. Mike it in several places to find this reading, then subtract it from maximum guide bore diameter to get maximum stem-to-guide clearance. Use the unworn stem diameter to determine what stem-to-guide clearance would be if new valves were installed.

How Much Clearance?—Typical stem-to-guide clearance is 0.001–0.003 in./0.02–0.07mm for both intakes and exhaust and 0.0015–0.004 in./0.03–0.10mm for just the exhausts in high performance engines or those with aluminum heads. Consider maximum allowable wear to be 0.005 in./0.12mm. Just remember when making your decision that the more a guide is worn, the faster it will wear. The amount of maximum stem-to-guide clearance you can live with should be determined by how many good miles you want from your engine after the rebuild.

For how to recondition valves, turn to page 91.

CYLINDER HEAD AIRFLOW

Before continuing on about reconditioning guides, I thought I'd take a moment to offer some irrefutable statements related to cylinder head airflow. First off, the horsepower produced by an engine at any rpm is directly related to the amount of air flowing into it provided the air/fuel mixture is correct and efficiently burned. All of the air entering an engine must go through the intake ports and all of the gases that leave must go through the exhaust ports. Sucking or blowing air through a cylinder head on a flow bench doesn't change the flow of a cylinder head.

The Bubble Bursts—Few other statements made about cylinder heads are as true as those above, but the following are close—and they are counter to many popularly held beliefs. Believe that: A cylinder head can flow too much air for a given engine. Indiscriminate reshaping/enlarging of cylinder-head ports usually reduces the flow of a head. Larger valves in a head can reduce flow. Highly polished ports flow less air than rough ground ports. A head that flows more air generally results in an engine that produces more power than one with higher compression. Other considerations have to do with port velocity and flow quality.

So how do you improve the flow efficiency of a cylinder head? This can be done only through the use of the right equipment, specifically porting tools in the hands of a professional and a flow bench where he checks his work. So if you plan on doing your own cylinder head modifications, confine your work to areas that "guarantee" improved airflow. We'll look at this work later but for now let's take a quick look at each of the points made above so you'll be wiser when it comes to modifying or choosing cylinder heads for your engine.

Stoichiometric—The best all around air/fuel ratio is called *stoichiometric,* which translates to 14.7 pounds of air for every one pound of fuel. For maximum power the air/fuel ratio will be as low as 12.5:1. If you consider that air weighs about 5 one-thousandths of a pound for each gallon at standard temperature and pressure and gasoline weighs about 6 pounds per gallon, 1200 gallons of air must flow into an engine for every gallon of gasoline. When you consider the engine must inhale at least 12.5 times the amount of air in weight of air for each pound of gasoline, that's a lot of volume.

Improving Port Flow

It's possible that an engine's cylinder heads can flow too

Only way to determine how much cylinder head flows is by testing on flow bench. Flow of ports is checked after making small modifications.

much air. Why? An engine needs to pump only so much air at any given rpm. With a given head, airflow velocity in the ports drop with engine displacement. This is why smaller carburetors and valves are used on smaller engines and low rpm engines—to maintain airflow velocity. Heads with too much flow capacity kills the bottom-end power of an engine for the same basic reason as does over-carbureting. Low intake velocity means poor fuel atomization, inefficient air/fuel mixture burning, reduced power at low-rpm, poor driveability and higher fuel consumption—all bad.

With that said, the fact is most engines usually don't have such a problem. This is particularly true if increased power at higher rpm is desired. It's then difficult to get too much air into an engine equipped with the original heads. So if you don't have a Boss 429 Ford, Hemi Chrysler or similar big-valve head in a street car engine, chances are your engine could benefit in the cylinder head flow department.

If you are determined to modify the heads but don't have a big bank account and a head porter with a flow bench, the best thing to do is restrict cylinder head modifications to areas that are most likely to improve flow. Otherwise take your heads to a professional who is in the business of flowing and porting cylinder heads. But if you insist on doing it yourself, keep in mind that indiscriminate removal of metal from a port usually reduces flow.

Areas to modify that will almost always improve airflow are the valve seats and valves. If done correctly, flow with the valves just off their seats will greatly increase. In fact, up to a 25% increase in airflow can be realized in the first 1/4 inch of valve lift. Not much will be gained from that point on, but the gain is major considering the work required to get it. Note that the following modifications described don't change port size. Consequently port velocity is not only maintained, it is increased due to the increased volume of air flowing through it.

Modify Seats—Do this with a three-angle valve seat similar to that described on page 91. Start by grinding or cutting the 45° valve seats so their outside diameters match the outside diameter of the intake and exhaust valves, respectively. Blue the seats and do a 30° top cut to establish seat OD. The top cut is made on the combustion-chamber side of the seat, or above the valve. It should be about 0.020-in. smaller than the valve OD so the valves will overhang their seats by at least 0.010 in. at the periphery. Under no circumstances should seat OD be larger than valve OD. Check seat diameter with a pair of dividers and a 6-inch scale.

To establish seat width, make a 60° or 70° bottom cut. This is made on the port side of the seat, or below the valve. Intake seat width should be no less than 0.060 in.; exhaust seat width should be no less than 0.080 in. Check seat widths with a small strip of heavy paper. A business card is a good thickness. Make two sets of tick marks at the edge measuring 0.080 and 0.060 in. between them, respectively. The bottom cut can be blended into the port with a die grinder.

Modify Valves—Do this by making a 30° back-cut on the back side of the valve head between the face and stem. All you have to do is break the sharp edge at the valve face. Be very careful that you don't get into the seat so far that the valve face won't have full contact with its seat. It should be at least 0.100 in. smaller than the seat.

Radius the margin into the underside of the valve head. On the exhaust valves, all you need to do is break the sharp corner. This is called the *clip angle*. As for the intake valves, radius the margin into the valve head.

Even though it's easier, you don't need a valve grinder for doing the work just described. If the valves are new or faces have been reconditioned, all you'll need is a drill press and a flat file or die grinder with a barrel-shaped stone. Chuck the valve in the drill press and turn it on. Use the flat file or die grinder to make the back cut and margin-to-valve head radius. Use a pair of dividers to check that you don't intrude on the seating portion of the valve face. Be careful that you don't touch the valve face with the file or grinder.

Manifold to Cylinder Head Matching

Although the following is not a port modification, it certainly affects port flow. It's also something you can do—match cylinder-head-port shape to those of the intake and exhaust manifolds.

The transition between the cylinder-head ports and intake or exhaust manifold runners should be smooth for efficient flow. It's OK, however, for the intake manifold runners to be slightly smaller than the intake ports at their mating faces, but not larger. The reverse is true at the exhaust side. It's OK for the exhaust manifold runners to be slightly larger. But under no circumstance should a gasket extend into the port. Its opening should be at the least the same shape as the shape of the port at its

Extensive modifications were made to intake and exhaust ports one step at a time in between data gathering on flow bench. Once desired intake and exhaust ports shapes were arrived at, remaining intake and exhaust ports were shaped to match, a time-consuming and expensive operation.

Cylinder Head Airflow, cont.

entrance or exit, or slightly larger.

The problem is checking port-to-manifold matching. This is the most difficult and important part of the port-matching process. For example, if yours is a V-type engine, you'll have to do a trial installation with the heads, intake manifold and gaskets on the engine, page 150. Otherwise you simply need to do a trial manifold installation to the head with the gasket in place.

With the manifold installed—intake or exhaust—match mark them. Do this with layout bluing, a straight edge and a scribe. Paint with bluing across the manifold, gasket and head at both ends of the manifold flanges. Once dry, lay the straight edge across the gasket joint 90° to it and scribe a straight line. The line should show up well on the manifold and head, but may be difficult to see on the gasket, so make a good indention when you cross it. And to ensure you don't lose track of which end of the gasket is which, offset the marks from one end of the manifold to the other. In other words, make the match mark at the upper end of the flange and the other at the lower end. Make sure you do both sides on a V-configured engine. Check the mark across the gasket and deepen it if it's not highly visible before you remove the manifold. Also, indicate on the gasket **MANIFOLD SIDE** and, if it's a V-type engine, **RIGHT** or **LEFT**. The gasket is the key to port matching. Remove the manifold.

To check port opening, lay the correct gasket to your match marks and tape it in position to the manifold gasket face. Use your match marks to align it. Check that the gasket doesn't hang over the port openings. Trim it to align with the edges of the port openings if it does hang over. Remove the gasket from the manifold and fit it to the head in the same manner, aligning it with your match marks. If the intake ports are larger than the manifold runners, all is OK. If they are smaller, blue the gasket face on the head. Then reposition the gasket to the head and, using the gasket as a template, scribe around the port entrances where they are smaller. You can do the same to the intake manifold, but the gain will be negligible at best. The reverse is opposite at the exhaust side of the head: The exhaust port exit should be smaller than the entrance to the manifold.

To enlarge a port entrance or runner exit, you'll need a pneumatic or electric die grinder with a 1/4-in. collet chuck and some stones and/or burrs. Burrs are sometimes called *carbide cutters*. Enlarge the port or runner opening by blending it gradually to the scribe line. This will smooth the transition into the head or exhaust manifold, thus eliminating the flow-killing sharp edge.

Recondition Guides

As I mentioned earlier, there is more than one way to recondition a valve guide. Some are better than others, but there are a lot of opinions as to which is best. If you don't like what one engine machinist says, ask another one. To cut through the smoke, following are the facts about valve-guide reconditioning as *I* know them. From these you can make your own judgments.

As I said earlier, different shops have different ways of doing guide work. Your challenge is to determine which way you would like your guides reconditioned, then find the shop that uses the method you've settled on. Whichever method you choose, be aware that guide work must come before valve seat work. This is to ensure that the seat is concentric with the guide. For the same reason, anytime valve guide work is done, the valve seats must be reconditioned.

Let's take a closer look at the most popular guide reconditioning methods in detail.

Knurling—This is by far the cheapest way to restore stem-to-guide clearance. The problem is a knurled valve guide is like anything else that's cheap—cost and quality go together. A knurled valve guide doesn't last long. You don't pay much, but you don't get much, either.

To knurl a guide, a special knurling tool is run into the guide. This tool doesn't add material to the guide. Instead, it raises thread-like ridges on the guide surface, effectively reducing the inside diameter of the guide bore. To arrive at the correct ID, a ream is run through the guide.

The problem with this method is only the tops of the ridges now support the valve stem, thus greatly reducing the guide-to-stem bearing area. It would be like putting skinny tires on a high-performance car. Wear is rapid even though perceived stem-to-guide clearance is restored. Oil consumption will be just as bad or worse than before. So don't knurl the guides if you want a top-notch job. On the other hand if a long service life is not a concern, knurl away.

Oversize Valve Stems—If you find that the valves need replacing because of stem or tip wear or anything else that makes them unserviceable, replacing them with ones that have oversize stems is an option. This approach is even more desirable if the guides need to be reconditioned. All

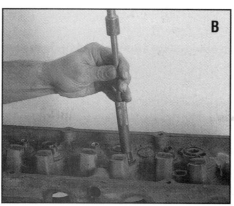

Installing thin-wall valve guide insert starts by reaming original guide oversize so it will accept insert. Insert is then driven into place using inner and outer mandrels. Excess material is trimmed away, then insert is expanded so new guide fits tightly in original guide. Guide is reamed to final size.

you need to do is install the new valves and restore the guides by reaming them oversize.

Before you take this approach, check the availability of valves with oversize stems for your engine. For example, standard, 0.003-in./0.07mm, 0.015-in./0.38mm and 0.030-in./0.76mm-oversize stems are usually offered. Before you make the final decision on which way to go, compare the combined cost of valves with oversize stems and reamed guides to the cost of new standard valves and reconditioned guides.

Valve Guide Inserts—In my opinion, using thin-wall inserts is the most cost-effective way of reconditioning valve guides. They can actually perform better than the original guides.

Thin-Wall Inserts come in two basic styles: A sleeve-type that is driven into place and a spring-like insert that is threaded into place. Material for both is a bronze alloy.

The thin-wall bronze valve guide insert is installed by first reaming the original guide oversize. The insert, which is approximately 0.060-in./1.52mm thick, is then driven into place using a special driver and expanded by running a spiraling tool through it. Because the insert is longer than the guide, excess material is trimmed and the guide is reamed to size.

To install the thread-type valve guide insert, a thread is tapped into the existing guide. After tapping, the insert is threaded into the guide and expanded to lock it into place so the valve and the insert can't work sideways in the thread, and so the backsides of the thread will be in intimate contact with the cylinder head for maximum valve stem-to-guide heat transfer. Finally, the guide is reamed.

Integral vs. Replaceable Guides—It's now time to distinguish between integral and replaceable valve guides. Integral valve guides are an integral part of a cylinder head. This is almost always the case with cast-iron heads, however some cast-iron heads use replaceable guides. The material used is either cast iron or a bronze alloy.

Replaceable guides are always used in aluminum heads simply because aluminum doesn't have the needed strength or durability. Whether it's a cast-iron or aluminum head, the guide is installed in a hole that's smaller than its OD—0.0015-in./0.04mm smaller is typical. This interference, or press fit, is what retains the guide in the head.

All of the previous mentioned guide-reconditioning procedures can be used to restore both integral and replaceable guides. However, an obvious way to recondition guides in a cylinder head with replaceable guides would be to replace the guides. But this method is not without its problems and pitfalls. First, all reference as to how far the guide is pressed in is lost. So you must first have the factory spec on this dimension or measure how far the top of the guide is above the spring seat and how far it projects into the valve pocket by referencing it to the valve seat, whichever is easier. If the replacement guide is the same length as the one being removed, you'll only need to measure it at one end or the other.

To remove and install replaceable guides, you'll have to know which way they are removed and installed and by what method. Although most replaceable guides come out from the top—the valve-cover side—and are pressed in from the top, you should

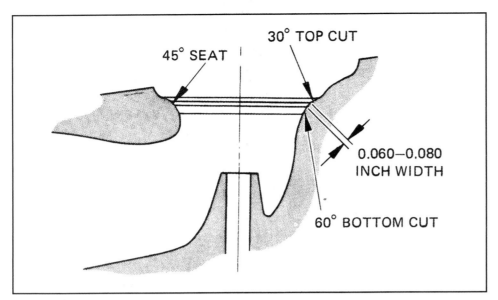

Valve seat is ground first—45° in this case—then outside diameter is established by 30° top cut and seat width is established by 60° bottom cut. Outside diameter of seat should be about 1/16-in. smaller than valve OD.

causes valve *shrouding*. This reduces flow through the port, thus reducing the power output of the engine.

Your best insurance against anything like the valves from being sunken is to use a good machine shop. If they have a good seat and guide machine, chances are the seat-and-guide work will be accurate. Here, the head is fixtured in a machine and the guide and valve machining is done with a rigid spindle. But the equipment is secondary. The skill of the machinist is your best insurance against bad engine work.

VALVE SEAT RECONDITIONING

Valve seats are reconditioned using special hand or power tools that turn cutters or grinding stones. With valve seats that have been reconditioned before or are damaged, you may not be able to get an acceptable seat. If this is the case, the old seats may have to be removed or bored out and valve-seat inserts installed in their place.

To repeat, both valve-seat reconditioning and replacement are jobs for an expert using the proper tools. The tools are too expensive for the home mechanic and the experience is required to do the job right. But it is useful to know about valve and seat work so you can talk intelligently to your machinist.

In addition to restoring valve seats, valve seat machining has another important benefit. It centers the valve seats to the reconditioned valve guides. This is because the grinder or cutter pilots off a shaft installed in the valve guide, making the seat concentric with the guide. However, if you didn't do any guide work and the valve seats are OK, seat work is unnecessary. If any guide work was done, the seat-to-guide relationship may have been changed and the seats

confirm this with the factory manual or an engine rebuilder. As for how to remove and install them, you'll need a stepped, or shoulder, punch and a brass or lead mallet.

The shoulder punch has a small diameter section that inserts into the guide for centering the punch and a larger section that shoulders against the end of the guide. With the punch in the guide, drive it out the top and drive in a new one.

Two hints for making guide installation easier: With machinist's blue and a scribe, scribe a line around the guide so it will align with a flat surface at the spring seat when it's in the correct amount. Secondly, chill the guide in a refrigerator immediately prior to installing it. This will reduce the guide's OD so it will drive in easier. Once in place, the guide is reamed to size.

Thick-wall inserts are similar to replaceable guides. The difference is thick-wall inserts are installed in cylinder heads that have integral valve guides. The difference here is the old guide is enlarged considerably more that it would be with thin-wall guides. The same installation method and press fit used for replaceable guides is used for thick-wall guides.

Replaceable valve guides come in two materials: cast iron and bronze alloy. Cast-iron inserts basically restore the guides to original condition. Bronze alloy guides are not only more durable, they are more expensive. Aluminum silicon bronze guides are normally good for well over 150,000 miles. They are easy on valve stems, too. If you want maximum durability, these are the ones.

A final word about valve guide work: If the new guide is way off center in relation to the valve seat, excessive valve seat grinding will be required. This will *sink* the valve—move it into the head—causing the stem to project farther out the top of the guide. This has three ill effects. It results in incorrect rocker-arm geometry—the rocker arm will not move across the tip correctly as it opens and closes the valve. A sunken valve will also increase combustion chamber volume, thus reducing compression. This may not sound so awful if you have a '60s to early '70s high compression engine, but it also

After pilot is firmly in place, valve seat grinding mandrel with stone is slipped over pilot. Mandrel and stone are rotated with pneumatic or electric tool.

must be reconditioned.

Valve seats are usually ground to a basic 45° angle. To ensure positive valve sealing a 1/2° or 1° *interference angle* is sometimes added, making the seat angle 45-1/2° or 46°, respectively. Although rare, intake-valve seats may be ground to a 30° angle. A quick look at your valves or valve seats will tell you whether a 45° or 30° seat is used.

Three-Angle Seat

A three-angle valve seat is the most cost-effective way of increasing air flow through a cylinder head. Top and bottom cuts of 30° and 60° or 70°, respectively, are made on both sides of the 45° valve seat. The top cut is made on the combustion chamber side seat and the bottom cut on the port side.

When reconditioning valve seats, the machinist makes the 45° seat cut. If you ask for a three-angle job, he then changes to a 30° stone or cutter and makes the top cut to establish valve seat OD. Seat OD is cut about 1/32-in./0.79mm smaller than the diameter of the valve face.

Next, the bottom cut is made with a 60° or 70° stone or cutter. This cut in the bowl side of the port establishes seat ID and width. This brings up the question, "How wide should a valve seat be?" The answer to this isn't simple. Different manufacturers specify different seat widths. For intakes, 0.050–0.100-in./1.2–2.5mm width is the range; for exhausts, seat width should the wider at 0.060–0.110-in./1.5–2.8mm.

How Wide?—How wide should the seats should be in your engine depends on what it's to be used for. If it will be used for performance only, ask a competent engine builder who knows the type of engines you're building. Generally, the narrower seat will flow better, but will be less durable. But if your engine is for towing or street use and it's durability you want, the valve seats should be on the wide side, particularly those for the exhaust valves.

Not only must the machinist have special equipment to grind or machine valve seats, he must also have equipment for measuring his work, particularly if he's doing a three-angle seat. To start with, he'll coat the new seat with machinist's blue. This makes it much easier to measure as cuts are being made. Seat widths and diameters are made with dividers and a 6-inch scale. Seat concentricity is made with a special dial indicator that pilots and rotates in the valve guide.

RECONDITIONING VALVES

Now that the valve guides and seats have been attended to, turn your attention to the valves. You should have already checked them for obvious damage such as burnt heads and excessive stem wear. If you haven't, do it now.

Dark bands are valve seats. After 45° seat is cut, seats are blued so they'll be contrasted from 30° and 60° cuts to ease measuring. Narrow strip of paper with tick marks, divider and 6-inch scale is all that's needed for measuring seats. Seat widths shown are 0.090 in. and 0.060 in. for exhausts and intakes, respectively.

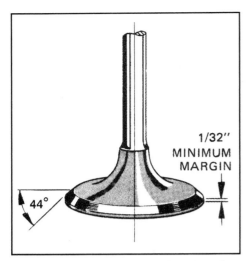

Exhaust valves should have margins of at least 1/32 in. after grinding to minimize risk of burning. Intake valve margins should not be less than 1/64 in.

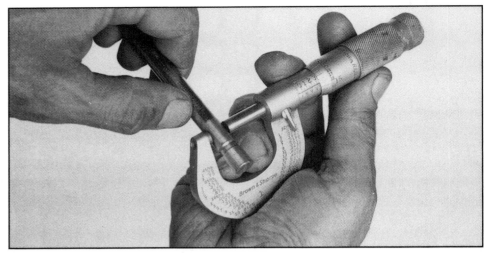

Using a 1-inch micrometer, compare measurements taken from an unworn section of valve stem and that having maximum wear. The difference between these two measurements is stem wear.

Valve Grinding

For valves that passed inspection, it's time to recondition them. The only acceptable way of doing this is by grinding, or resurfacing, the tip and valve face—the portion of the valve that contacts the seat when closed.

The tip end of the valve stem is ground first because it centers the valve in the machine so the face will be ground true to the stem. If the tip is untrue, the face will also be untrue.

The valve stem is clamped in a fixture at one end of the grinding machine so the tip can be passed squarely across the face of a rotating grinding wheel. Only enough material is removed to remove any irregularities. Remove too much and you risk going through the hardened valve tip surface and upsetting rocker arm geometry. Because of this, major damage such as mushrooming renders the valve unserviceable. Discard it. After the tip is resurfaced, it is chamfered to remove the sharp edge and burr.

Valve faces are reconditioned by grinding off small amounts of material to the correct angle until all flaws are gone. This is done by chucking the valve stem end first in a spindle. Its face is then rotated against and across a rotating grinding wheel, ensuring the face is true. As the valve is ground, fluid is directed on the valve for cooling and to wash away metal and abrasive particles.

Margin—After grinding a minimum amount of material from the valve face, check to see that there is sufficient margin—thickness of the valve head at its outer edge. Insufficient margin, especially on an exhaust valve, may cause it to burn.

As for margin width, the minimum should be larger for a larger valve. For example, a 1.75-in/44.4mm exhaust valve should have no less than a 0.040-in./1.0mm margin and a 1.40-

Valve grinding starts by resurfacing and chamfering the valve stem tip. The valve face is then ground at an angle that's the same as or 1° less than the valve seat. I prefer grinding it at the same angle.

in./35.5mm exhaust valve should have 0.020-in./0.5mm minimum margin. Typically, though, exhaust valve margins should be at least 0.030 in./0.8mm. As for the margin on an intake valve, its minimum can be half that of the exhaust valve because of its cooler operating environment. Therefore, the typical minimum intake valve margin is 0.015 in./0.4mm.

Face Angle—If your valves are ground at a 30° or 45° angle and a 1° interference angle is needed, grind the valve face angles to 29° and 44°, respectively. I prefer grinding the valve faces to the same angle as the seats.

After grinding the valve tips and faces, your valves should be ready to go back in the heads. If all matching surfaces are restored—seats, guides, faces and tips—there's no need to keep the valves in order. But if they are OK and didn't need reconditioning, keep and install them in order.

Lapping Valves—You may have heard the term *lapping* valves and wonder where this operation fits into the valve and seat reconditioning processes. The truth is if a valve is worn bad enough that grinding is required, hand-lapping won't correct it. And if a valve and seat are ground correctly, lapping is not necessary. Consequently, lapping is not used in commercial valve work, but racers and specialists do it regularly as a check.

If you lap your valves, you'll need lapping compound, or paste, and a lapping stick. The lapping paste, which is an abrasive, is applied to the valve face. The lapping stick has suction cups on both ends for holding the valve as you rotate it. With paste on the valve face, install the valve in its guide and apply the lapping stick the valve head. Rotate the valve back and forth lightly against its seat by moving the palms and fingers of your hands back and forth against the stick.

After lapping in a valve, be sure to clean off all lapping compound. Abrasives and engine parts are not compatible. The gray band on the valve face and seat from lapping should be full circle, indicating full contact between the valve and seat. If it's not full circle on the valve, the valve face is not concentric with the stem. If it's not full circle on the seat, the seat is not concentric with the guide. As you finish lapping the valves, organize them in such a way so you'll install them on the seats they were lapped on.

Do They Seal?—If you want to be sure the valves seal before you run your engine, there's an easy way to do

Lapping a valve starts by coating valve face with lapping paste. Valve is then installed on its seat and oscillated with light pressure using a lapping stick between the palms of your hands. Valves should be installed on the seats they were lapped on, so keep them organized.

it. You'll first have to install the valves with the springs either temporarily or permanently. Do it temporarily and you won't have to install the valve stem seals, spring shims or inner springs. To install the valves permanently, read on for how to install seals, check springs and whatever.

Either way, position the head so the combustion chambers are facing up and the head gasket surface is level. Using kerosene, fill each combustion chamber and let the head sit a few hours, preferably overnight. If any chamber empties or is lower after the time period and a port is wet, you know a valve leaked. If the intake valve was the culprit, the intake port will be wet and vice versa. You'll have to redo that valve or seat. If you don't

Lapped versus unlapped valve. Gray band on face should be full-circle to ensure sealing. If it's not, face is not concentric with stem and valve should be reground. Same goes for valve seat. If it's not full-circle, face is not centered on stem.

think there is any leakage, check the ports anyway. Using a flashlight for illumination, peer into each port. They should be dry. If you find a leaky valve, check with your engine machinist on remedying the problem.

VALVE SPRINGS

Next on your list of things to inspect are the valve springs. Their condition is critical to proper engine operation, particularly an engine that will be run at higher rpm ranges. Certain inspection steps must be performed to verify that the springs meet minimum standards. Those that don't cannot be reconditioned. They must be replaced.

To appreciate the work a valve spring does, it must close the valve against the inertia of the rocker arm, pushrod, lifter, valve, valve retainer and a portion of the spring itself. At an engine speed of 3000 rpm, it must do this 25 times a second! Double this to 50 times a second at 6000 rpm! And it must do this while operating in a hostile environment, particularly the exhaust valve spring.

The cam lobe raises the lifter and pushrod to operate the rocker arm and compress the valve spring as it opens the valve. But the cam lobe doesn't close the valve. The spring must do this. It must push everything back in place as it keeps the lifter in contact with the closing side of the cam lobe. Loss of control of the oscillating mass called the valvetrain and the engine experiences what is called valve float. And when valves float they don't close, resulting in a severe to total loss in power. If hydraulic lifters are used, they pump up, further aggravating the problem. Worse yet a floating valve may contact the piston, resulting in valvetrain and piston damage. So check those springs.

Valve Spring Terms

To inspect valve springs, it is necessary to have a valve spring tester of some kind. And to understand what's involved in valve spring testing, you must understand a few basic terms and appreciate the importance of maintaining minimum standards. Common terms related to valve springs are: spring rate, free height, load at installed height, load at open height and load at solid height.

Spring Rate—is not a commonly listed valve spring specification, but it relates directly to most of the other specs. It is, however, one of the basic terms necessary to describe a coil spring's mechanical properties.

Spring rate, or stiffness, governs the load exerted by a spring when it is compressed a given amount. Rate is usually expressed in so many pounds per inch (lb/in.) of deflection. That is, when a spring is compressed a specific amount, it requires a specific force to do so. For example, if the rate of a valve spring is 400 lb/in., it will exert an additional 200 pounds when compressed 0.500 inch. Remember, the spring was compressed during installation, so total force, or "tension," is 200 pounds plus whatever the initial force was at the installed height.

 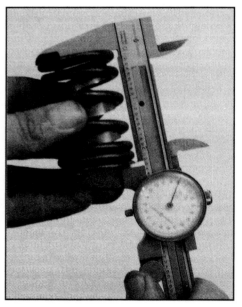

Two quick ways of determining whether a valve spring needs shimming and/or replacing. First, spring should be square. One shown is being checked with a carpenter's framing square. A valve spring should lean no more than 1/16 in. at the top. Check free height too. If heights vary from each other as much as these two check their load, or pressure, at installed height. Final check is to measure free height directly. Note flat wound wire inside spring. It's a damper.

Engine designers use the lowest possible valve-spring rates, even for racing engines. This is because stiffer springs require more power to turn the camshaft and operate the valvetrain. Stiffer springs also cause unnecessary valvetrain wear. So only enough "spring" should be used to close the valves at the engine's redline—maximum rpm limit.

Typical valve-spring rates vary between 300 and 450 lb/in. Spring rates are usually indicative of the rpm range of the engine and the weight of the valvetrain. For example, as engine speed increases, so must spring rate. The same goes for increased valvetrain weight, which translates to increased *inertia*—resistance of a mass to a change in motion. Lighter titanium valves and aluminum or titanium spring retainers reduce inertia; heavier roller lifters increase inertia.

Increased spring force is required to close a valve, or keep it from floating at high rpm. If a valve spring loses its rate, or resiliency, it is no longer capable of closing the valve at higher rpm ranges. Spring fatigue and the resulting loss of spring tension is caused by the many thousands of opening and closing cycles. Additionally, spring material is softened—*annealed*—by heat transferred to the base of the exhaust valve springs from the exhaust port. Spring fatigue first shows up in a loss of free height. Remember this when checking springs, especially those for the exhaust valves. That brings up a point: spring order. You should've kept the springs in order.

Free Height—This is how tall a valve spring is when sitting uncompressed. If a spring's free height is less than specified, then too will be the closing force it exerts on the valve. This is simply because the spring will be compressed less as installed. Therefore, understanding spring rate and free height gives you a clue as to why a spring won't meet its load specifications.

Load at Installed Height—This is the force a spring should exert on the spring retainer when the valve is on its seat, or is closed. Installed height is the height measured from where the spring sits on the cylinder head to the underside of the spring retainer.

Note: Frequently, the term *pressure* instead of *force* or *load* is incorrectly used when spring specifications are discussed, so don't be confused.

A typical load-at-installed-height specification is 75–85 lb. at 1.810 inches. When this spring is compressed to 1.810 inches, it should exert a force of between 75 and 85 pounds. The absolute minimum installed load is 10% less than the lower limit, or 68 pounds. For the common "grocery getter," a spring that meets this minimum is OK. But for a high-rpm engine, your valve springs should be at the upper end of the specification.

Load at Open Height—Another common valve spring specification, this is the force a spring exerts on the retainer when compressed to the height it would be when the valve is fully opened. A typical specification is 240–260 lbs. at 1.330 inches. Similar to the installed-height load, a spring must fall within the 10% minimum force limit. Otherwise, it should be replaced. In this case the minimum is 216 pounds. Again, consider what your engine is to be used for when checking the springs.

Solid Height—The solid height of a coil spring is measured when the spring is compressed to the point where each coil touches the adjacent coil. The spring at this height is said to go *solid*, or *coil bind*.

Coil springs should never be compressed to this height in normal

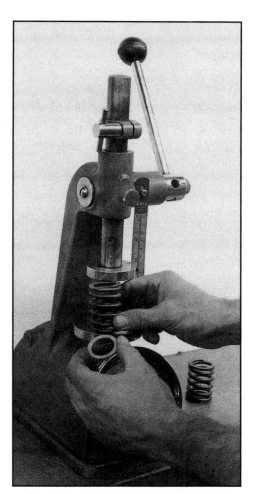

Checking spring load at installed height. Adding shims can be used to bring load within specification, but coil clearance should be checked with spring on shims at open height to ensure it doesn't bind, or go solid. There should be at least 0.060 in. (1.5mm) between coils. You have a choice of 0.015-in., 0.030-in. and 0.060-in. shims.

service. If a valve spring were to reach its solid height before the valve is fully opened, the load on the valvetrain would cause the weakest link in it to fail. More often than not, a pushrod bends.

Squareness—The squareness of a valve spring is how straight it sits on a flat surface. This is important so the spring won't force the valve stem sideways against the guide, causing uneven and excess guide and stem wear. Limit valve spring out of square to 1/16 in. measured from the top of the spring to a square with the spring sitting on a horizontal surface.

Spring testers used in machine shops are very expensive, however this one is relatively inexpensive. All you'll need is a bench vise and 6-inch scale. Gauge reads out directly in pounds.

Checking Valve Springs

Now that your head is spinning with spring terms bouncing around in there, let's look at how to check valve springs. The main things to check for are squareness, free height, installed load, open load and binding.

To make these checks, one thing you'll need are the valve spring specifications for your engine. If you are rebuilding the engine to production specifications, you'll need the manufacturer's specs. Otherwise, you'll need the cam manufacturer's specs. If two valve springs are used—an inner and outer—check them separately. If a flat-wound damper is used, remove it when checking loads.

Unless you have an engine machine shop or are among the fortunate few, you probably don't have access to a spring tester. To say the professional type shown is expensive would be an understatement. Its cost puts it out of the do-it-yourselfer budget.

Fortunately, there is a valve spring tester that is affordable. It's not as easy to use, but you can get the same information as you would get with the high-priced tester. All you'll need is the tester, a vise and dial calipers or a 6-in. scale. Clamp the tester with the spring on top of it and compress the spring to the specified height. Read spring force directly. Measure installed or open height with the calipers or scale.

If your local auto parts dealer or speed shop doesn't sell such a tester, contact: Goodsen, 4500 West 6th Street, Winona, MN 55987; or C-2 Sales & Service, Box 70, Selma, OR 97539.

Squareness—This is easy to check, so let's start with it. A carpenter's framing square or anything that can check a 90° angle will do. Stand the spring on the horizontal surface or leg of the square and position it against the vertical surface. Turn the spring until it leans away from vertical so the gap at the top coil is widest. Limit this gap to 1/16 in.–0.625 in./1.6mm. If it exceeds this, replace the spring.

Free Height—This will give you a quick check of a valve spring's condition. A spring that has sagged—has gotten shorter—by more than 0.100 in./2.5mm should be replaced or checked with a spring tester. It's not unusual that the exhaust-valve springs will have sagged because of being overheated—heat is the number-1 enemy of a valve spring.

A good rule of thumb is to replace any spring that is shorter by 1/8 in. (0.125 in./3.2mm) or more. A spring that is short by 1/16 in. (0.0625 in./1.6mm) can be used for light duty, but not for high-rpm use.

If you've replaced some springs or have purchased all new ones, you should still check them. Put them in the lineup and perform the following tests. If you didn't keep the springs in order, from now on you must do so. Spring inspection from this point on requires that you use a spring tester. If you don't have one, take the springs to your engine machinist for checking.

Springs that fail the free-height test by 1/16 in. or less can be shimmed to their original height. Shims are available at auto parts stores for adjusting spring height and loads. VSI has them in various thicknesses; 0.015 in./0.4mm, 0.030 in./0.8mm and 0.060/1.5mm being the most common. You'll only need a 6-in. scale, vise and feeler gauges to check free height and for coil bind at open height. In the vise, compress the spring to the open height and check for a 0.012-in./0.3mm minimum clearance between coils.

If you use a shim, wire it to the spring. And don't forget to keep the springs in order.

Load—Checking installed load and open load requires a spring tester. But if all springs fell within the free-height specification and the engine is not for high performance use, chances are the springs are OK to use. You can get by without further checks. But if they needed shimming or you simply don't want to leave things to chance, a spring tester must be used.

On the question of shimming, there are two reasons it is better to replace a spring than to shim it. Fatigue or annealing caused the spring to sag and shimming won't correct either condition. The spring will continue to sag. Second, shimming used to correct load will compress the spring more, increasing the rate of fatigue. Rapid loss of installed and open loads will result.

So unless you're doing a patch job on your engine—knurled valve guides and pistons and minimums throughout—don't shim to obtain installed and open loads. Replace the springs. But if you do shim, remember to wire the shim to the spring and keep them in order.

A final point about shims: Use only one shim per spring. If you shimmed a spring for free height then need to shim it for load, forget the one used for free height correcting and shim only for loads. Finish your spring inspection by checking for coil bind.

ASSEMBLE CYLINDER HEADS

To assemble the cylinder heads you'll need a valve-spring compressor, a 6-inch scale and maybe some shims. I say maybe because you probably won't need them unless you did extensive valve and seat work.

Installed Height

The distance between where the spring sits on the head and bottom side of the spring retainer when a valve is on its seat is actual installed height of the valve spring. Because the force exerted by a spring is directly related to how much it is compressed—remember spring rate?—it's important that installed height is correct.

Installed height increases as material is removed from a valve seat and valve face during reconditioning. This causes the valve stem to project farther out of the valve guide which moves the spring retainer the same distance from the spring seat. If this distance is more than it should be, spring loads will be less and valve float will occur at lower engine speeds.

Adjusting installed height is also done by shimming. Don't get this confused with shimming for free height and loads at installed and open heights. Shimming for installed height reduces the distance between the spring seat and retainer to achieve the specified installed height.

Check Installed Height—Do this by installing the valve in its guide without the spring. If a rotator, steel spring seat or whatever should install between the spring and head, install it on the spring seat. Oil the valve stem and install the valve in its guide.

Again, oil the upper end of the valve stem and install the valve stem seal using the plastic sleeve installer. If not, use a strip of clear tape over the valve stem tip. The advantage of installing the seal at this point is it helps hold the valve in place.

In case you installed positive-sealing type seals or other special seals, follow the manufacturer's directions for installing the valves and seals. If installed height checks out OK, you won't have to remove the valve. The job of installing this valve and seal is finished.

Install the spring retainer with its keepers. Pull up on the retainer to seat the keeper and measure the distance between the underside of the retainer and top of the spring seat or whatever you have sitting on it. For measuring you can use one of two methods: If you have a telescoping gauge, use it and a 1—2-in. micrometer. If not, a 6-inch scale or vernier caliper will do. Record your measurements for each valve. Subtract specified installed height from your measurements to find whether a shim is needed. Anything less can be ignored. For more than 0.015 in./0.4mm, choose the nearest shim that's thinner. An easier way I found to measure installed height is with a gauge made from heavy wire or welding rod. Cut a section of wire longer than installed height, then file or grind it to the specified installed height. Square both ends of your "gauge." Check the length with 1–2-in. mikes or vernier caliper.

Shim Thickness—To determine shim thickness needed, insert the gauge between the rotator or valve seat. If the gauge doesn't fit snugly, use feeler gauges between the top end of the wire gauge and spring retainer to determine shim thickness. Again, use the shim with a thickness nearest to, but thinner than what is needed to obtain exact installed height. To repeat, if the difference is 0.015 in./0.4mm or less, a shim is not needed.

Record the thickness of the shim on a small strip of paper and slip it under the spring that needs that shim. Do this for each spring. Also, record this information in your shopping list. When you return from the auto-parts store, you can easily match the shims to the correct springs.

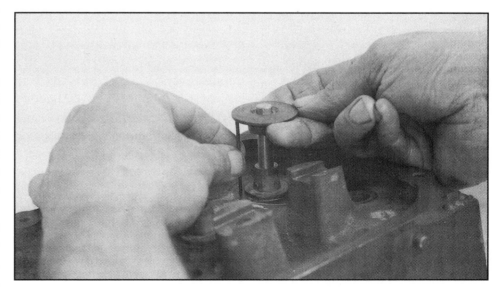

I found that a short section of heavy wire cut and filed to match spring installed height made checking seat-to-retainer distance very easy. If distance is excessive, install shim and recheck using wire "gauge." When you find the right combination, organize valves and shims so you can install them in the positions in which you checked them.

PERFORMANCE RETAINERS & KEEPERS

Standard retainers and keepers are good for anything but all-out racing. If you use high-rate springs with stock retainers and operate the engine at high rpm, they may not be sufficient. The valves can pull through the keepers or a retainer or keeper may break.

Aftermarket manufacturers offer high strength steel, aluminum and titanium retainers. If the engine will be used for street driving and an occasional high-rpm blast, steel retainers are best. They cost less, wear better than aluminum or titanium retainers and are less apt to crack or break.

Another feature with high performance keepers is a larger angle on the keepers. As an example, 10° is used rather than 8°. Be careful when switching from one to the other that you don't interchange keepers. An 8° keeper won't work with a 10° retainer.

Reserve aluminum and titanium spring retainers for use only if you will do frequent and thorough inspections. When doing teardowns, use the keepers that match the retainers and replace them at every rebuild.

Before you check the next valve, make sure the retainer has sufficient clearance to the seal when the valve is fully open before you make the following check.

Retainer Clearance

The next thing you need to check is the clearance between the retainers, oil seals and guides. There shouldn't be any problem in this department unless you will be installing a high lift camshaft. The bottom of the retainer should clear the oil seal by at least 0.060 in./1.5mm at maximum valve lift. This is simple to do, make the check regardless of whether you think there will be a problem.

While holding the retainer in place, open the valve to its fully open position and push the seal down on top of the guide. Check valve opening with a 6-in. scale. This is easier with a telescoping gauge set to spring open height or a wire gauge that matches spring open height. Eyeball the retainer-to-guide or seal clearance. Have your feeler gauges set to measure 0.060 in./1.5mm just in case the retainer appears to be too close.

If the retainer is too close to the seal or guide, the top of the guide must be shortened. Have your machine shop do this. This is usually a problem only encountered when a high lift camshaft is installed, but a newly installed guide that's too long will cause the same problem.

After you've checked retainer-to-seal clearance, remove the retainer and keepers and place them with the spring(s) and shim(s) for this valve.

Note: If you will be changing to a high performance dual valve spring, the inside spring may not clear the umbrella-type valve stem seal. Check to make sure by referring to the manufacturer's instructions. If clearance is a problem, you'll have to install positive-sealing Teflon seals. This requires machining the tops of the valve guides to a smaller diameter so they'll accept the Teflon seals. Most engine machine shops can perform this simple operation for you.

Assemble Cylinder Heads

You should now have everything needed to assemble your cylinder heads, tools included. With the parts organized and laid out accordingly,

Valve, spring, Teflon seal and related hardware organized for installation into race head for Grand National NASCAR engine. Organization is important regardless of engine, be it for racing or the street. Titanium retainer and high-angle keepers are used. Spring assembly is made up of inner and outer valve springs with flat wound damper in between. Shims and seat install between spring and aluminum head.

you should be able to assemble the heads quickly and without problems.

You may find that more than one shim goes with one spring because you shimmed for free height, loads and installed height. There could also be a separate spring seat or rotator for some or all of the springs. Then again, the spring may sit directly on the head. Just make sure you have all the necessary hardware that goes under the springs.

Position the heads on the bench with the valves, seals and, if they are used, separate spring seats or rotators. Don't forget that the springs and shims must be installed just as they were checked—with *their* valves. To do the cylinder head assembly job, you'll also need a squirt can of oil, a spring compressor and some grease. A plastic mallet, lead hammer or brass hammer will also come in handy.

When installing a shim, note that

After oiling valve stem and guide, slip valve into its guide. Umbrella type seals install over valve stem. Softer polyacrylic seal is the easiest to install. Push seal down over stem until it bottoms against guide. It will hold valve closed, freeing up your hands so you can install the spring(s), retainer, keepers and any shims or spring seats. Note springs and shims over valves lined up in order. This will help eliminate installation errors.

the serrated side goes down, or toward the head. When installing springs, check for coils at one end spaced closer than at the other. If this design is used, the end with tighter coil spacing installs against the head. And don't forget to install the internal springs or dampers if they are used.

Compress a valve spring only enough that you can install the keepers. Once the spring is compressed, install the keepers, too. A little grease on the keeper grooves will hold the keepers in place while you release the spring compressor. Check that the keepers are secure by tapping the tip of the valve with a soft hammer. This will also help seat the keepers.

Installation of the rocker arms will complete this job, but this is done after the heads are on the engine. It's a good idea, though, that you inspect the rocker arms and recondition or replace them now to prevent delays during engine assembly.

ROCKER ARM ASSEMBLIES

Factory rocker arms are either forged or stamped steel. And whether your engine uses shaft-mounted rocker arms or individually mounted ones, the two areas to check are the same: the rocker arm pivots and valve tip ends. As for the pushrod sockets, they "never" have a problem. Give each a quick look anyway just to make sure. You should also inspect the pivot for wear. It will be individual ball or barrel pivots, or a shaft that all rocker arms pivot on.

Individually Mounted Rocker Arms

For this type of rocker arm, remember to keep the pivots and rocker arms mated. If you didn't, replace them all.

Check the pivot and rocker arm interfaces on both the rocker arm and pivot. If you find scratches and gall marks or serious wear in these areas, replace them. At the valve tip end of the rocker arm, look for serious wear. If it's not too bad, the tip end can usually be reconditioned by grinding on the same machine used for grinding valves. But if the valve-tip end is grooved, replace the rocker arm.

One type of rocker arm cannot be reconditioned in this manner. Rail rocker arms like those used by Ford on some engines built from the late '60s through the late '70s must be replaced if wear is excessive. The

Spring seats and/or shims, if used, are installed between spring and cylinder head. In some cases, a shim or seat won't fit over the valve stem seal, so place them on spring pad before you install seal. Once seal, seat and/or shim are in place, install spring(s).

Compressing spring no more than necessary, install keepers. Grease on keepers will hold them in place while you release compressor. Using soft mallet, tap valve stem tip to check that retainers are firmly seated in grooves.

flanges that straddle the valve tip to stabilize, or guide, the rocker arm also prohibit the reshaping of the valve tip mating surface.

Once you've inspected and reconditioned or replaced the rocker arms and pivots, store them in preparation for reassembly. Store them in pairs where they will be free from dust and moisture. Include the attaching hardware and any baffles that may install with the rocker arms.

Shaft-Mounted Rocker Arms

Reconditioning this type of rocker arm assembly can be a major operation. Although wear occurs in the same locations as with the individually mounted rockers, disassembling shaft mounted rockers can be a chore.

To remove rocker arms, you must slide them off the shaft. But before you begin, note the order in which the rocker arms and other components assemble such as springs, washers, attaching hardware and mounting stands. Also note the orientation of each shaft and rocker arm. As with the early Chrysler Hemi, there may be two shafts for one head, one for the exhaust valves and one the intakes In this case, five stands supported both shafts. Finally, the rocker arms may be offset, or angled, because the pushrods and valves operate in different planes. A line drawn through the centers of the pushrod socket and valve tip end will cross the rocker arm shaft at an angle other than 90°.

Don't depend on your memory. If your engine's rocker arm setup is as complex as most are, lay each rocker arm and shaft assembly on a plain background and photograph them. You'll be glad you did come assembly time. Take several shots at different angles. One shot should be of the top of the rocker arms and another the bottom.

Disassembly—This can be nearly impossible if your engine has accumulated many miles. This is due to rocker arm shafts being coated with deposits such as varnish. To minimize this problem remove the deposits by soaking the assemblies in carburetor cleaner or a strong solvent such as lacquer thinner. Use a stiff bristle brush to scrub off the varnish. Dry the assembly with compressed air then coat it with penetrating oil. The following work will be easier if the parts are at least room temperature, or 70 F (20 C).

Start your teardown by removing any retainers such as pins, springs, spacers or washers from the ends of the shaft. Remember to keep everything in order. If a rocker arm or stand—if used—is stubborn, tap it off with a rubber or plastic mallet. Aluminum stands are the most difficult to remove. They are also the easiest to break, so be careful. String the parts on a wire as you remove them, then run the wire through a hole in the shaft after all parts are off. This will allow you to inspect mating wear surfaces on the rocker arms and shaft and also reinstall the rockers in their original positions.

If the rocker arms are oiled from the shaft instead of the pushrods, both ends of the shaft will be plugged. If yours uses this setup, remove the plugs so the shaft bore can be cleaned. A cup-type plug can be removed from one end with a sheet metal screw and Vise Grips after drilling the plug. Before pulling it out, note how far the plug installs in the shaft. Drive out the other plug with a long metal rod and hammer.

If the rocker arm shafts are really dirty, chances are they will be worn badly too, particularly on their bottom sides. Serious shaft wear will be apparent. It will be easy to see, let alone feel. Drag a fingernail along the bottom of the shaft and it will catch on the edge of the worn areas. Any roughness usually means excess wear. Mike the suspected areas of the shaft to be sure. Replace the shaft if there's more than 0.003-in./0.07mm difference between a worn and unworn section. If a shaft is worn out, replace it.

Finish cleaning the rocker arms, shafts and other components. Oil flow to the rocker arms can be blocked by deposits in the oil holes, so make sure they are clean and unrestricted. A large gun bore brush works well for cleaning inside the shafts.

Measure Clearance—Check the rocker arm to shaft clearance using a 0–1-in./0–25mm outside mike and a telescoping gauge. Measure rocker-arm-bore ID and the mating diameter on the shaft and subtract the two. Clearance should be 0.003–0.006 in./0.07–0.15mm for the standard rebuild and 0.0015–0.0030 in./0.04–0.07mm for high-perfor-mance use. To obtain a tighter fit, it may require

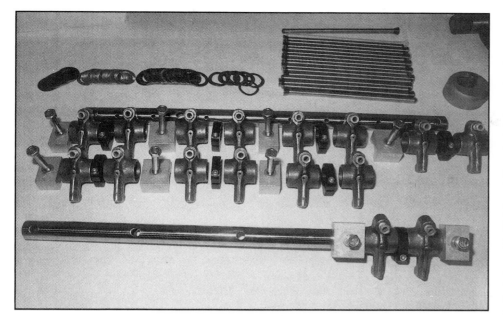

Shaft-type rocker arm parts are assembled in the exact position in which they were removed. Rubber mallet is used to lightly tap aluminum stand into place. Shaft with rocker arms, springs, stands, spring washers, flat washers and retaining pins on it make up this assembly.

selective fitting and reaming. This means you may have to install bushings in the rocker arms, an expensive proposition.

Before you replace a complete rocker arm shaft assembly, check with your engine machinist. Oversize shafts may be available. This will allow you to reuse the worn rocker arms. These and the stands, if used, are reamed oversize to fit the new shaft.

If valve lash is adjustable, each rocker arm will have an adjusting screw at the pushrod end. The screw will have either a locknut to maintain adjustment or a screw with an interference fit in the threaded portion of the rocker arm. If it's the latter type, check that it takes at least 7 ft-lb to turn the screw. You certainly shouldn't be able to turn it with your fingers. If it's less, replace it so lash will be maintained.

Assembly—Assemble the rocker shaft now that you've inspected and declared all parts usable as is or you have reconditioned or replaced the defective ones. Here's where organization pays off; photos, notes you've taken, keeping parts in order and maintaining cleanliness are all part of this. As with disassembly, it's best that you do the assembly at room temperature.

Assemble each rocker arm shaft separately. Have some assembly lube on hand such as 50W oil in a squirt can, Ford's Oil Conditioner or GM's EOS (Engine Oil Supplement). Use plastic garbage bags to cover parts.

Once lubrication is used, it's more important than ever to shield engine components from dust and dirt contamination.

Start the assembly by installing plugs in the ends of each shaft. For cup-type plugs, use a punch or old bolt ground flat on the end that's slightly smaller than the plug ID. Wipe a thin film of RTV sealer on the plug OD and drive them in square to the correct depth. For threaded plugs, install them with sealer on the threads.

Assemble one shaft at a time. Position the shaft, rocker arms and other components on the bench as you would view them as installed on the head. Refer to your notes and photos. If used, begin by lubricating the shaft and installing a rocker arm stand in the center position on the shaft. Make sure the notch or notches and oil holes in the shaft are oriented correctly to the stand. If the fit is tight, warm the stand in an oven.

Next to install may be a rocker arm followed by a spring or spacer, then another rocker arm, a stand and so forth. Whatever the order, be careful when installing offset rocker arms. They must be oriented correctly. To

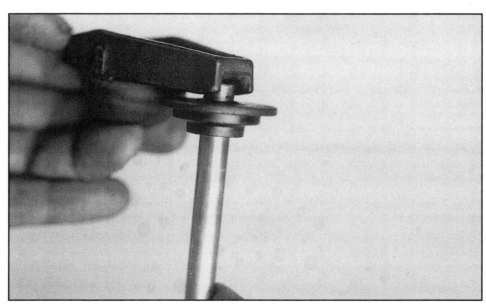

Infamous rail rocker arm, flanges on rocker arm straddle valve tip to stabilize rocker arm. If yours is this type, you can't recondition them. Replace them.

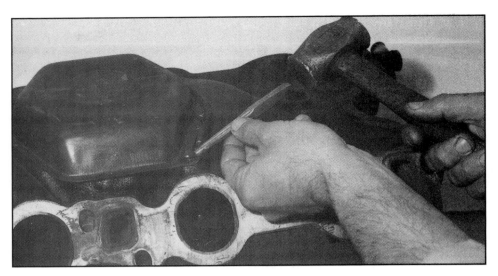

Don't overlook servicing bottom of intake manifold. If yours has a baffle attached to bottom it will have to come off so you can remove baked-on deposits. If rivets attach baffle, remove them by first raising each rivet slightly. Do this by lightly wedging small chisel under rivet head. This will allow you to clamp on rivet head with Vise Grip pliers. This done, carefully work out each rivet.

After cleaning underside of intake manifold, baffle is installed with rivets driven back into place.

be sure, refer to your notes and photos as you go. When you are finished assembling the rocker arm shafts, store them in a plastic bag. This will keep the rocker arms clean until the time comes for installing them on the engine.

INTAKE MANIFOLD

There's not much to reconditioning an intake manifold. About all you need to do is clean it. Such is the beauty of non-moving parts. But if yours is a V-type engine and you had the heads milled, both the sides and bottom of the manifold may have to be milled—if you haven't already done so. Refer back to page 81 for details on the relationship between cylinder head and intake manifold machining.

The easiest way to clean a cast-iron manifold is to have it hot-tanked or baked. Before you have this done, remove any plastic or aluminum parts or fittings first. If a divorced choke is used, remove the bimetal spring that installs in the heat-riser passage. The same goes for an EGR valve.

To clean an aluminum intake manifold, use a stiff-bristle brush and solvent combined with a lot of old-fashioned elbow grease. To really spiff up an aluminum manifold, have it bead blasted and sealed with a clear coat.

If a baffle is attached to the bottom of the manifold, remove it. This will expose the caked-on deposits so you can scrape them from the underside of the exhaust-crossover passage. The baffle will be retained with bolts or rivets. Bolts are easy enough to remove, but rivets take special attention. Remove them by carefully driving a very small chisel under the rivet head. This should back the rivet out enough to get hold of it with Vise Grip pliers. With the pliers clamped on the rivet head, carefully unscrew it while prying underneath the plier jaws.

Heavy deposits can be removed from water and heat-riser passages with a blunt screwdriver and a round file. The gasket surfaces should be perfectly clean. To ensure they are, run a large flat file across their surfaces.

After cleaning the manifold, reinstall all the components you removed from it. Seal the threads of anything that installs in the water jacket such the water temperature sender or ported vacuum switch. If used, don't forget the bimetal choke spring or EGR valve. Use a new gasket under the EGR valve. Reinstall the heat-riser baffle, too. If rivets are used, start them in their holes and tap them into place with a small hammer. If you broke a rivet or simply don't like them, use a 1/4-20 tap to thread the hole and use bolts with lock washers to retain the baffle.

COMPRESSION RATIO

The theoretical compression ratio of an engine is determined two factors; displacement and combustion chamber volume. More simply, it's the volume above the piston when it's at BDC compared to the volume above it at TDC. In mathematical terms it looks like this:

CR = (SV + CV)/CV

Reduced to a simpler form it looks like this:

CR = SV/CV + 1

where:
CR = compression ratio
SV = swept volume (displacement) of one cylinder.
CV = combustion chamber volume, or clearance volume.

To see how this works, let's determine the compression ratio of a 350 CID V8 that has a 85cc combustion-chamber volume. The first thing that must be done is to get displacement and combustion-chamber volume in the same terms. Because it's easier and more accurate to work in bigger numbers I convert displacement from cubic inches to cc's. There are approximately 61 cubic inches per liter and there are 1,000cc's per liter. Making this simpler, there are 16.39cc/in.^{3}. The displacement of one cylinder in cubic centimeters is: 350 in.^{3} X 16.39cc/in.^{3} = 5738cc; 5738cc/8 cylinders = 718cc. Compression ratio can now be calculated by plugging in the numbers:

CR = SV/CV+ 1
CR = 718cc/85cc + 1
CR = 9.44:1

These numbers are fine and dandy, but the problem is getting them in the first place. Displacement is certainly no problem. It's relatively easy to measure or look up bore and stroke and do the calculation. Getting clearance volume (CV), though, can be difficult. Be aware that clearance volume is not only made up of the volume in the cylinder head. It is also affected by the shape of the piston (dished, flat or domed), where the piston stops at TDC in relation to the block deck (deck clearance) and the installed thickness of the head gasket. The shape of the piston can add or subtract from combustion-chamber volume depending on whether it is dished or domed.

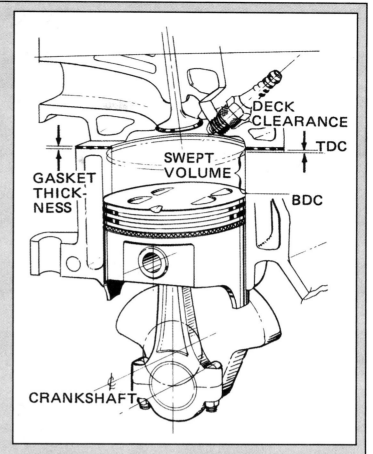

Compression ratio is determined by swept volume and clearance volume of an engine. Swept volume is that displaced by the piston as it moves between TDC and BDC. Displaced volume of one cylinder is simply engine displacement divided by the number of cylinders. Clearance volume is the volume contained between the top of the piston at TDC and cylinder-head combustion chamber.

Other factors can complicate the situation. For example, the 409 Chevy has all of its combustion chamber in the block. Milling 409 heads has negligible affect on compression ratio. At the other extreme, a domed piston with zero deck clearance such as is the usual case with a Hemi-head engine has all of its clearance volume in the head less the volume displaced by the piston dome.

Swept Volume

To determine compression ratio, the first thing you'll need is the swept volume (SV), or displacement, of one cylinder. To find SV, plug bore (D) and stroke (S) into the following formula:

Swept Volume = $0.7854 D^2 \times S$

Clearance Volume

This volume (CV) is not so easy to determine because

of the irregular shape of the typical combustion chamber and piston top. Rather than determining CV through calculations alone, you'll have to do some measuring using the *cc'ing* process. This involves filling cavities with liquid and determining their volumes in cubic centimeters (cc's), thus the term cc'ing. Why cubic centimeters? Because scientific burettes, the instrument used for measuring, are calibrated in cc's.

The process for determining clearance volume is not so difficult as it is cumbersome. For this reason I'll run through the process using a real-life situation. For making your calculations, just substitute the numbers. Even though your numbers will be different—larger or smaller—they'll be similar. The important thing to realize is all machine work must be done. The valves and spark plugs must be installed in the head. For measuring clearance volume in the block, a rod and piston fitted with rings must be installed.

The example I use is an engine that displaced 454 cubic inches that's been bored 0.030-in. oversize. Bore and stroke are now 4.280 in. x 4.000 in., respectively. Displacement is now 460 cubic inches.

Swept volume (SV) is the first thing to determine. Using the same formula shown above:

$$SV = 0.7854\ D^2 S = 0.7854 \times (4.280\ in.)^2 \times (4.000)$$
$$SV = 57.55\ in.^3$$

To convert cubic inches to cc's, multiply swept volume in cubic inches by 16.39.

$$SV\ in\ cc's = (57.55\ in.^3)(16.39\ cc/in.^3) = 943cc.$$

Clearance Volume (CV) is much more difficult to determine than SV. It requires more than making a simple calculation. You must determine several volumes through a combination of calculations and actual measurements.

Combustion-Chamber Volume (V_h)—This is the first component of clearance volume to measure, which is done by cc'ing the chambers. To do this, you'll need some tools and equipment. First on the list is the graduated burette. Get a plastic one if possible. Next is a 6-in. square or bigger piece of clear acrylic plastic that's at least 1/2-in. thick. Drill a 1/4-in. hole through the plate within 2 in. of one of its edges. Countersink one end of the hole with a larger drill to make filling easier and leave the other end of the hole square, just deburr it. The countersunk side will be up.

You'll also need some measuring liquid, too. A quart of automatic transmission fluid mixed in equal amounts with clear cleaning solvent works well. The red ATF makes it visible for reading graduations and the solvent reduces its viscosity. Kerosene also works well, but is not as visible.

Finally, you'll need some thin grease such as Lubriplate for sealing. To measure piston travel, get a dial indicator and a clamp-on or magnetic base. The indicator must have as least a 1-in. travel.

The valves and spark plug must be installed in the chamber you're measuring. Position the head upside down so the gasket surface is not quite level, but a little higher on one side than the other. Smear a light film of grease around the combustion chamber on the gasket surface. Don't apply too much.

Position the plastic plate against the head so it covers the combustion chamber. The chamfered side of the fill hole should be up and to the high side of the combustion chamber. This will help in filling the chamber and purging any air bubbles. Push down on the plate and work it back and forth, sealing it to the head. To prevent an inaccurate volume reading, make sure grease doesn't squeeze out from under the plate and into the combustion chamber. If it does, remove the plate and reseal it. Now you're ready for the burette.

Close the petcock and fill the burette to the zero mark with measuring liquid. If it's over-full, drain the fluid until it's at the zero level. The bottom of the *meniscus*—part of the liquid that climbs up the inside of the burette—should align with the zero division at the top of the graduated scale. You'll need a burette that has a scale reading from zero at the top so you'll know how much fluid has been drained from the burette into the cavity you're checking. Remember to always read from the bottom of the meniscus.

To fill the combustion chamber, hold the tip of the burette squarely over the fill hole in the plate and open the petcock. Hold it so fluid goes straight into the combustion chamber without hitting the edge of the hole. If there's not enough fluid to fill the chamber, shut the petcock so the meniscus aligns with the bottom division and note the volume. In this example the combustion chamber volume was over 100cc, so I drained the fluid down to the 100cc division. Refill the burette as before and finish filling the combustion chamber. In this case an additional 19cc completely filled the combustion chamber.

Check for leaks around the valves and spark plug. If you find any leakage under the plate, in the ports or

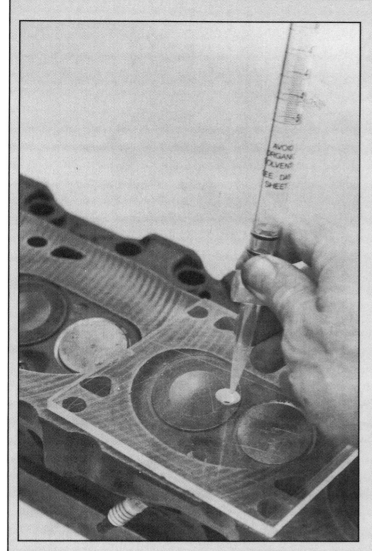

Checking cylinder head combustion chamber volume by filling with measure volume of liquid. Acrylic plate is sealed to cylinder head with light grease, then is filled with clean solvent, kerosene or alcohol mixed in a 50/50 ratio with automatic transmission fluid for color. Use red food coloring to make clear liquid easier to see.

around the plug, correct the cause and start over. If you don't, your results will be incorrect. Also check for air bubbles. Carefully tilt the head so the air bubble moves to the fill hole and escapes.

Total combustion-chamber volume is the sum of the two burette readings, or in this example:

$V_h = 100cc + 19cc = 119cc.$

To do the job right, cc each combustion chamber and record the results. You'll then be able to equalize them to the largest chamber. This is done by removing material with a cutter or grinder from the smallest chambers to bring their volumes up to that of the largest chamber.

The next component of clearance volume (V_c) is found in the cylinder. It may add to or subtract from the clearance volume, depending on the shape of the piston. This volume and that affected by piston deck clearance must be compensated for to find the volume added or displaced by the piston top shape. Finding V_c requires a combination of measuring and calculating. This is because the volume is measured above the piston with it positioned an exact distance down the bore. In the example I used 1 inch. Half that amount is OK, but 1 inch makes calculations easier even though it requires the use of more measuring liquid.

Let's start with the measuring part since you're in that mindset. Just as when cc'ing the head, you'll need the burette, plate, measuring fluid and light grease. Additionally, you'll need a 1-inch travel dial indicator with a magnetic base.

Use the same process that was used for cc'ing the heads. If your block is on an engine stand, position it so the deck surface is up, but at a slight angle. Lock the stand or prop the engine up in this position. Run the piston about 1-1/2 inches down the bore. To keep the measuring fluid from leaking past the rings, coat the cylinder wall with light grease immediately above the piston, then run the piston up to within 1 in. of TDC.

Use a dial indicator for measuring. Set the indicator dial to read its 1-in. maximum travel with the plunger on the piston at TDC. Rotate the crank to bring the piston up so it's exactly 1 in. below TDC. Run the piston up until the dial reads 1 inch as measured to a flat section of the piston.

Position the plastic plate on the block deck, sealing it with light grease. Locate the plate so the filling hole is near the edge of the high side of the bore and the chamfered end of the hole is up. Again, watch for grease that may squeeze out from under the plate. Fill the burette, aligning the bottom of the meniscus to the zero graduation. Using the same method you used when cc'ing the combustion chamber, open the petcock and drain fluid into the cylinder. It can take more than two full burettes and then some to fill the cylinder. In the example it took 194cc to fill the bore, just 6cc short of two complete 100cc burettes full, or:

$V_{p1} = 194cc.$

There are three calculations for volume in the block to make: V_{p2}, V_g and V_d. V_{p2} compensates for moving the piston 1 in. down the bore; V_d allows for deck clearance; and V_g allows for head gasket thickness. Taking them in order, make the calculation for V_{p2} using the same equation you used for Determining SV. Plug in actual bore size and use 1 in. for stroke. In the example:

$$V_{p2} = 0.7854 \times (4.280 \text{ in.})^2 \times (1.000 \text{ in.}) \times (16.39 \text{cc/in.}^3)$$
$$V_{p2} = 236\text{cc.}$$

To get the volume affected by a piston dome, dish or valve reliefs, use the following equation. Plug in the numbers just calculated for Vp1 and Vp2:

$$V_p = V_{p1} - V_{p2}$$
$$V_p = 194\text{cc} - 236\text{cc}$$
$$V_p = -42\text{cc.}$$

Note that V_p is negative. This is because the domed piston in the example displaces clearance volume. As such, this piston reduces the space—clearance volume—in which the air/fuel mixture is compressed, thus resulting in higher compression. Had the number been positive, as would be the case if the piston were dished, the opposite effect would result—more clearance volume and lower compression.

Deck Clearance (V_d)—This volume calculation supplies the volume needed to compensate for the position of the piston relative to the block deck. If the piston is below deck at TDC, it adds to clearance volume; if it's above deck, Vd subtracts from CV; zero deck clearance has no effect. Before you can make this calculation you'll need to determine deck clearance. Do this with a dial indicator supported by a magnetic base.

Set up the indicator so the magnet is on the block deck, the indicator is square to the piston and reaches the center of it if there's a flat area for the plunger to rest on. This will avoid any error introduced by the piston "rocking" in the bore. Once you have the stand adjusted, shut off the magnet, swing the indicator over the deck surface and zero the dial. If the piston is domed, such as with the example piston, measure on a flat area that's to either side of the dome, but directly over the wrist pin. If there's no flat at either side of the dome, measure deck height at two locations: square to the wrist pin on flat surfaces to the extreme left and at the same distance from the center of the piston to the extreme right. To minimize any error from a "rocking" piston, add the results and divide the sum by two to get deck height. In the example deck height is 0.010 in. below deck. In this case volume V_d adds to clearance volume:

$$V_d = 0.7854 \times (4.280 \text{ in.})^2 \times (0.010 \text{ in.}) \times (16.39 \text{cc/in.}^3)$$
$$V_d = 2.36\text{cc}$$

To obtain clearance volume from the head gasket, two dimensions are needed; installed gasket thickness and inside diameter of the gasket fire ring. Gasket thickness must be determined from a gasket that has been installed, or compressed by the force of correctly torqued head bolts. The problem here is the gasket springs back—partially returns to its original thickness—when pressure from the torqued bolts is removed. This prevents you from getting an accurate reading by installing the gasket, then removing it and measuring it. So you must do one of two things: Find an engine like yours that's assembled with the same head gasket and measure the gap between an overhanging part of the head and deck surface. Make the measurement with the inside portion of vernier calipers or, if that's not possible, do it with feeler gauges. The second method is to place a small piece of lead between the head and block deck where there's no gasket and install the head and torque the bolts to spec. This will compress the lead to the same thickness as the gasket, but the lead won't spring back. Therefore, the thickness accurately represents the installed thickness of the head gasket. In the example, head gasket thickness was 0.035 in. The second dimension you'll need is the inside diameter of the fire ring. The problem here is the fire ring may not be a perfect circle. If this is the case, estimate its ID by measuring it in several locations. Any error from your estimate will be so small it will have little effect on your compression ratio calculation. In the example I estimated head gasket fire-ring ID to be 4.300 inches. Clearance volume due to the head gasket is:

$$V_g = 0.7854 \times (4.300 \text{ in.})^2 \times (0.035 \text{ in.}) \times (16.39 \text{cc/in.}^3)$$
$$V_g = 8.33\text{cc}$$

Summarizing, separate volumes and their values making up clearance volume in the example engine are:

V_h (Cylinder head combustion chamber volume) = 119cc
V_p (Volume due to shape of piston top) = -42cc
V_d (Volume due to deck clearance of piston) = 2.36cc
V_g (Volume due to thickness of head gasket) = 8.33cc

Total Clearance Volume (CV) = $V_h + V_p + V_d + V_g$
CV = 119cc - 42cc + 2.36cc + 8.33cc
CV = 88cc

Compression Ratio (CR) = CV + SV/CV
CR = 88cc + 943cc/88cc
CR = 1031cc/88cc
CR = 11.7:1

Increasing Compression Ratio

You now know how to determine the compression of your engine. But you may want to increase its compression to the maximum allowed by rules governing your racing class or that permitted by the octane of available gasoline. The question then becomes how much do you reduce clearance volume to obtained the desired compression? Again, it's a combination of math and measuring. Given the desired compression, you can easily determine the new clearance volume. Putting the revised compression-ratio formula to work, let's assume the desired compression for the example engine is 12.5:1. Using the revised formula and plugging in the numbers, the new clearance volume is:

CV_2 = SV/(CR_2 — 1)
CV_2 = 943cc/(12.5 — 1)
CV_2 = 943cc/11.5:1
CV_2 = 82cc

To increase compression from 11.7:1 to 12.5:1 in the example engine, the clearance volume change is CV_2 — CV_1 = 82cc — 88cc = -6cc. The negative number indicates CV is reduced by 6cc. Do this by milling the cylinder head to reduce combustion chamber volume 6cc. The new combustion chamber volume V_h will be 119cc — 6cc = 113cc.

Determining Milling Amount—But by how much should a cylinder head be milled to obtain the desire volume? This is the first question. The second concerns valve-to-piston clearance. Removing material from the bottom of a head not only reduces chamber volume, it reduces valve-to-piston clearance by nearly the same amount. To check this clearance, turn to page 138. For now, though, let's look at how to determine how much needs to be milled from a head to arrive at the desired combustion chamber volume.

Start by setting up the head similar to the way you would for cc'ing it. There are some major differences in how to do it this time. First, the head must be level from end to end and from side to side. Also, you don't need the clear plastic plate for doing this procedure. As for the burette, do just as before: Fill it with measuring liquid to the zero graduation. This time, though, drain the fluid into the combustion chamber to the new volume, or 113cc for the example engine. The distance from the gasket surface to the liquid surface is the amount you need to have milled from the head. Note: The smaller the combustion chamber, the more you must remove from the head. In the example, 0.029 in. had to come off the head.

Measure from the head-gasket surface to the liquid surface using a dial indicator or the depth gauge end of your vernier caliper. Be aware, though, that the capillary action of the liquid causes a similar effect as to what creates the meniscus in the burette. It will "jump" up to meet the indicator plunger or caliper, causing a lower reading. Exactly how much this difference is depends on the liquid being used, but 0.010 in. is typical. So add 0.010 in. to your reading, or in the example the measurement was 0.019 in. With 0.010 in. added, 0.029 in. is the amount to mill off. As a final check, cc the head after it's milled. ■

Complete cylinder head reconditioning by giving gasket surface a light cut on milling machine. Note polished combustion chambers, which was part of porting job. Cylinder head was tested on flow bench to ensure modifications increased port flow.

ENGINE ASSEMBLY 6

From this point on, your engine project should be all down hill. Everything is clean, the parts are new or reconditioned and most of the hard work is behind you. It's time to reassemble your engine!

Before you get started, keep the following points in mind. Be very careful when assembling your engine. This is your last chance to discover any problems such as incorrect clearances or parts. And never leave internal engine parts or a partially assembled engine uncovered.

ASSEMBLY SUPPLIES

Just like the work you've done up until now, there are a few things you'll need for assembling your engine. Have a large, clean, garbage bag handy. When taking a break or when you finish for the day, slip the bag over the engine or parts. Dirt and moisture are an engine's worst enemy. The bag will protect it and not leave lint as would a cloth rag. You'll also need various types of lubricants, sealers, a complete gasket set, core plugs and your favorite color engine paint. It's best that you get these before you start. Finally, there are some special tools you'll need as I detail later in this chapter.

Engine built for street use, occasional car shows and some racing. Compromise was made between an engine for the street but powerful enough for competition. Compromise favored racing, so driveability for street use was sacrificed.

Lubricants

There are three functions assembly lubricant plays in a newly rebuilt engine: The obvious and most critical is lubricating internal engine parts during those first few minutes of initial start-up. The second one is corrosion protection. If your engine won't be run for an extended period of time for whatever reason, keeping those freshly machined surfaces rust-free is critical. The third function is bore sealing. Without oil to seal the rings to the bores and pistons, your engine won't have sufficient compression for that initial start-up.

One of the liquid lubricants to use for assembling your engine is the oil you'll be using in it. You might as well get a case of it. While on the subject of oil, make sure you use a graded detergent oil in your engine. This information is on the can or jug. Use the best grade of oil you can find, or an SH or higher rated oil. As for weight and brand, only in politics and religion will you get more opinions. Generally, though, 30W oil is fine. But I prefer a multi-viscosity oil such as 5W-30. It acts like 5W oil when it's at 0F and 30W oil when it's at 212F. Oils with wider ranges of viscosity such as 10W40 are not recommended by auto manufacturers. As for brand, that's up to you. It's hard to go wrong in this department if you use a name brand and follow the recommendations I've just made.

As recommended in the previous chapter, Ford's Oil Conditioner or GM's Engine Oil Supplement (EOS) make excellent assembly lubricants. They are good for lubricating bearings, bearing journals, rocker arms, lifters, oil pumps and pushrods during assembly. Another choice for lubricating these items is Lubriplate's No. 105 lithium grease. For lubricating high-load areas such as rocker-arm tips, rod, main cap and cylinder-head bolt and stud threads and the undersides of bolt and nut heads, use an extreme-pressure lubricant such as CMD #3 from Chicago Manufacturing and Distributing.

Moly Lube—Another lubricant that stands above all is molybdenum disulfide, or simply moly. I wouldn't use anything else for prelubing cam lobes and lifters. It is great for cam and lifter break-in during the first few minutes of engine running. It is so good, in fact, that many aftermarket high-performance camshaft manufacturers supply it with their camshafts in either grease or liquid form.

Anti-Seize—The last lubricant is difficult to categorize. Anti-seize compound acts like a lubricant and sealer, but it's main purpose is to prevent threaded fasteners from, well, seizing. Exhaust manifold/header bolts are prime examples. The bolts frequently break off rather than unthreading. And anything that threads into aluminum presents a special problem, especially spark plugs. When installed without anti-seize compound on the threads, a spark plug will frequently "pull" the threads from the cylinder head when removed. So add anti-seize to your inventory, especially if your engine uses aluminum heads.

Sealers

Nearly every gasket in your engine will require some type of sealant. But each one has a different function, so different sealers must be used. The one-sealer-does-all doesn't exist.

RTV Sealer—If there ever was a sealer that's close to being universal it's RTV (room temperature vulcanizing) silicone. But it is available in different types. There's the "universal" type, high temperature RTV and one for use on O_2 sensors.

One thing RTV silicone doesn't do well is hold gaskets in place as they are installed. It acts more like a lubricant, making a gasket react like a watermelon seed that squirts out between your fingers. This is what frequently happens as the manifold, valve cover or whatever is tightened down. Otherwise it's a great sealer, sometimes suitable in place of a gasket! When buying an RTV silicone, make sure it's for automotive applications. Bathroom caulk won't do.

What works best to hold a gasket in place is 3M's Weatherstrip Adhesive®. Other names, many of which aren't flattering, have been given to this yellow, smelly and gooey adhesive/sealer, but they are descriptive. You'll get the picture after you've used it or have removed a

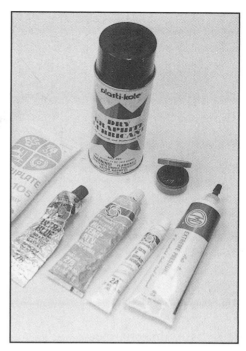

Products you'll need sooner or later; those to prelube and seal engine during assembly. Might as well round them up now.

gasket that's been glued in place with weatherstrip adhesive.

Examples of other sealers are Outboard Marine Corporation's Adhesive Type M and Ford's Gasket and Seal Contact Adhesive. Other excellent sealers include High Tack®, Coppercoat® and Gasgacinch®. These are all good for their intended applications. Sometimes the choice comes down to availability or your preference, nothing else.

Add a roll of Teflon® tape to your inventory. It is useful for sealing pipe-threaded components such as oil pressure senders, water temperature senders, ported vacuum switches and threaded oil gallery plugs.

Gaskets, Seals & Plugs

Providing a positive seal for gasoline, oil, coolant, combustion chamber pressures and manifold vacuum is a critical part of engine assembly. The sealants mentioned above accomplish part of the task, but the lion's share of engine sealing belongs to gaskets, seals and plugs.

Clean, organized assembly area is essential. A plastic leaf bag keeps your engine clean when it's not being worked on.

You'll get the gaskets in an engine rebuild set. If a complete set is not available for your engine, you'll need two gasket sets: a lower, or conversion set, and a valve grind set. As for the plugs, you'll need to get them separately in a set for your engine.

If you didn't rebore your engine, you can save money by purchasing a re-ring set. It not only has the gaskets, a re-ring set also includes the rings and bearings. Another option is a complete engine rebuilding kit. It comes with pistons of popular oversizes, rings, bearings, camshaft, lifters, plugs and gaskets.

Tools

Other than the tools normally occupying your toolbox as well as the ones I discussed in Chapter 2, you'll need some special ones for assembling your engine. You may be able to rent some, others you'll have to buy or, if you're lucky, borrow.

Ring Expanders—These are used for one job only—to spread compression rings so they can be installed on pistons. They are inexpensive and can prevent the expense of a complete new set of rings because you broke one. They can also save you a lot of physical pain. Your thumbs will be sore and bloodied if you use them to spread the rings.

A ring expander looks like a small pair of pliers. But rather than closing when you squeeze the handles, the jaws open. Easy to use, one hand uses the expander to spread, or expand, the ring and the other controls the ring as you fit it over the piston and into its groove.

Ring Compressors—These are used to compress rings on their pistons so they can be installed into the bores. In this case, there is no option: A ring compressor is a must for installing pistons. Of the three ring compressors, there are the cylinder, clamp and cone types. They all work well if used correctly. The cylinder type is most popular do-it-yourselfer ring compressor because it's easy to use and relatively inexpensive. The ring compressor preferred by most commercial engine builders is the clamp type because it's faster. Cone types are also fast, but one cone is required for each exact bore size. I recommend you use the cylinder or clamp type.

Plastigage—This is a handy bearing clearance checking tool even though experienced engine builders turn their nose up at it. Although you've measured the bearing journals and have ordered the appropriate bearing inserts, you shouldn't believe what you read on boxes. Oil clearances should be checked to make sure the right bearings got in the right box. The quickest and easiest way of doing this is with Plastigage, a small precision strip of colored wax. Purchase one sleeve of the green variety. It has a 0.001–0.003-in./0.002–0.008mm range. It should be fresh, not hardened by age. If it has been hardened from sitting on the shelf for a long time, your readings won't be accurate.

Cam Bearing Installer—This is a tool you should be acquainted with if you removed the cam bearings. If you removed them, you must decide on how to install them. The first may be the best approach. Take your block to the shop and have your engine machinist do it. But if you'd rather do it yourself, you'll have to purchase, borrow or possibly rent a cam bearing installation kit. If you decide to do it yourself, I tell you how later in this chapter.

Engine Stands—These are great for assembling engines. They are relatively inexpensive and make the engine assembly process easier. It's very easy to reposition the engine to improve access to any side, top or bottom. If you don't have one, now is a good time to think about getting one.

Assembly Area

Where you assemble your engine is an important consideration. An unlit shed with a leaky roof and a dirt floor isn't a good choice. You need a clean room. Lighting should be good, you should be reasonably comfortable,

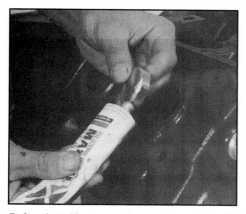 

Before installing core plug, apply sealer to hole ID or plug OD. Place plug over hole, then drive in plug using large diameter punch so flange is even with bottom of hole chamfer. You shouldn't be able to "catch" backside of plug flange with your fingernail.

and the area should be dry and dust free. There should also be adequate shelf and table-top space to allow for storage, organization and a clear work area. Now, let's get to work.

INSTALL PLUGS

Once the plugs are installed in the rear of the block, you can mount the block on an engine stand. These include the cam plug and rear oil-gallery plugs. Read on for installing these.

Camshaft Plug

Two basic types of camshaft plugs are used: cup and expansion. The cup plug used to seal the rear cam-bearing bore is similar to core plugs, but typically with shorter flanges. Expansion plugs looks similar, but are installed differently.

The cup plug installs with the flanges away from the cam. Run a small bead of silicone sealer around the ID of the bore before you begin installing the plug. Drive it in with a punch slightly smaller than the inside diameter of the cup. A large socket and a lead or brass mallet will work. Do not strike the edge of the plug flange. If the plug cocks as it's going in, either remove it and start over or favor the side that's sticking out as you continue driving it in. Drive in the plug until the edge of the flange is either even with or slightly past the bottom edge of the chamfer by no more than 1/32 in. If you can catch the back side of the plug flange with your fingernail, the plug isn't in far enough.

The expansion plug installs with the flange pointing in. To drive it in you'll need a hollow thin-wall punch that has an OD similar to that of the plug. Don't drive against the center of the plug. It may not seal if you do. A pipe reducer or short section of pipe will work as a driver. Apply sealer to the outside surface of the flange and drive in the plug until it bottoms in the counterbore or is flush with the bore, whichever comes first.

Core Plugs

Your engine will have core plugs in the sides of the water jacket and possibly the front and rear of the block. Cup plugs are used most of the time, but some high-performance engines use large-diameter pipe plugs.

Cup-type core plugs are similar to cam plugs, but some are easier to install because of their smaller diameters and longer flanges. They have less tendency to cock as they're driven in. As with cam plugs, you'll need sealer and a punch that fits loosely inside the plug. After applying sealer to the bore, drive in the plug until it's even with or slightly below the bottom of the chamfer. Check it with your fingernail to see if it's in far

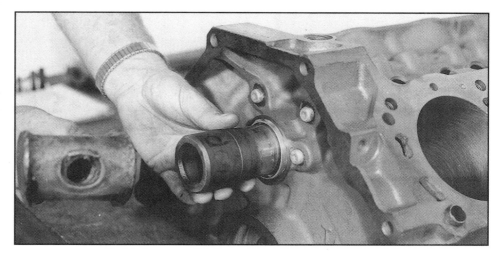

Cup-type cam plug is similar to core plug but has shorter flanges. After applying sealer to flanges, install in a manner similar to how you installed core plugs. Plug that has no flange installs with convex—open–side in. To secure it, drive center in until flat.

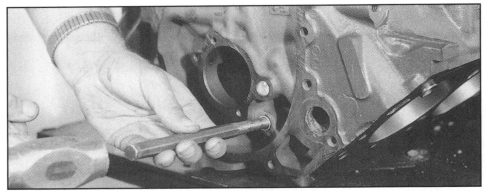

Cup-type oil-gallery plug is driven in with punch. Punch should not fit tightly in plug, otherwise flange will collapse on punch as plug goes in. Seal plug and drive in just below hole opening. Secure plug by staking flange in three places with small chisel.

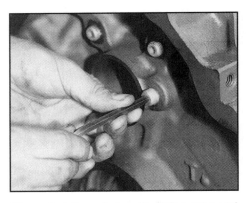

Wrap oil gallery plugs with Teflon tape and install. Leave tape off leading two threads to keep tape from breaking off and getting into oil passage. Firmly tighten plug.

enough.

If large-diameter pipe plugs are used, seal their threads and run them in tight. Use Teflon® tape for sealing. Wrap the tape counterclockwise around the threads as you would look at the plug when installed. Hold tension on the tape and pull it into the threads as you rotate the plug. If you removed the waterjacket drain plugs, seal and install them in the same manner.

Oil Gallery Plugs

Typically, pipe plugs install in the rear of the oil galleries and cup plugs in the front. To install the threaded plugs, seal them with Teflon® tape. Wrap the tape into the threads as just described. Be careful not to let the tape extend past the first thread. This could cause serious problems if the tape broke off and restricted an oil

SECURE THOSE PLUGS

If you're building a high-performance engine, secure the core plugs with rivets or epoxy and the oil-gallery plugs by staking them. Although it's unlikely that a correctly installed plug will come out, making sure they don't is good insurance.

Core Plugs—When securing core plugs I prefer using rivets instead of epoxy. Although it takes more effort, rivets are more positive and not as messy. An advantage epoxy has over rivets is it not only retains the plugs, it also helps seal.

You'll need at least three rivets for each core plug and a few spares. These are available at any industrial supply house. Get a sack of #00-1/4-in. long U-drive rivets. To prepare the block for the rivets, drill three evenly spaced holes adjacent to each core plug flange with a no. 54 drill. Start this by bluing the core plug hole with machinist's blue, indicating the rivet pattern and center punch them so each is 120° apart. Drill the holes square to the chamfer and so the rivet head will overlap the plug flange. Don't drill any deeper than necessary. To get the depth right, wrap a piece of masking tape or 200 mph tape around the drill back from its tip so this distance is equal to the length of the rivet plus 1/16 in. A rivet laid alongside the drill is a quick way of transferring this measurement.

Run the drill in until the edge of the tape is even with

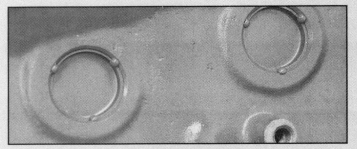

Racers pin core plugs so they stay put. Small brad-type rivets can be purchased at most fastener supply stores. Drill three holes square to chamfer at edge of each plug. Hole should be slightly smaller than rivet. Rivet heads should overlap plug flange. Drive in rivets to secure plugs.

the top of the hole and it will be at the correct depth. To be sure, install a rivet by lightly tapping it into place. If it pulls down tightly against the plug flange, great. If not, pull out the rivet, reposition the tape on the drill, drill the hole deeper and recheck it. Once you have all rivet holes drilled, install the rivets.

To epoxy the plugs in place, run a bead around the edge of each of the core plug flanges. The surface to be epoxied should be clean and oil-free.

Oil Gallery Plugs—Cup plugs sealing the oil galleries can be secured by staking. This is a simple process of distorting the leading edge of the holes with a chisel. Using a pattern similar to that used for the rivets, notch each hole in three equally spaced places. This raises the metal at each notch, preventing the plugs from coming out.

Check that cam bearings are the ones for your engine. Cross-check bearing part numbers to list on slip of paper in or on box, as is the case here.

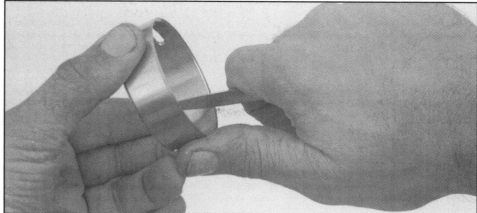
New cam bearings have square or ragged edges. To ease cam installation, chamfer edges of bearings with a pocket knife or bearing scraper before you install them.

passage or orifice, causing oil starvation to a critical component such as a hydraulic lifter or bearing.

CAM BEARINGS

If you plan on using an engine stand during engine assembly, don't mount the block on it just yet. The stand limits access to the rear of the block, so you'll need to install the cam bearings and rear plugs with the block on a bench or the floor. Once the plugs and bearings are in place, you can install the block on the stand.

Take a look at the cam bearing inserts before you begin installing them. Make sure they are the correct ones. Fit them to the cam-bearing journals if necessary. You can't Plastigage cam bearings.

Progressively sized cam bearing inserts are sometimes used—they get larger from the back of the block to the front. Other engines use cam bearings of the same size. But there could be other differences such as the oil holes and grooves. Most cam bearings have one oil hole except maybe for the front or rear one. One is the oil feed hole. It provides a passage for oil to flow to the bearing journal. The other hole may be for allowing oil to flow to the distributor shaft. Oil for the lifters or distributor shaft may flow through a groove in the insert. Refer to your notes on this one.

Chamfer Bearings

Prepare each bearing insert for installation by removing the sharp edge from both ends of the bearing. This will allow the cam to slide into place without hanging up. A bearing scraper works best for doing this. It looks like a triangular file without teeth and three hollow ground sharp edges. If you don't have one a pocket knife will work.

Chamfer each bearing edge by holding the scraper or knife at 45° to the edge of the insert ID and peel a shaving off about the width of four human hairs. Hold the bearing so you can rotate it as you chamfer it. Remove just enough material so you can't feel a burr when you drag a fingernail across either edge of the bearing.

There's a simple trick to align the oil holes in the inserts with those in the block. Draw a line with a felt marker across the back of each insert in line with the oil hole(s). Similarly, draw a line on the face of each bearing bore in line with oil hole(s) in the block. Align the marks on the bearings with those on the block prior to driving in the bearings to make sure the oil holes align.

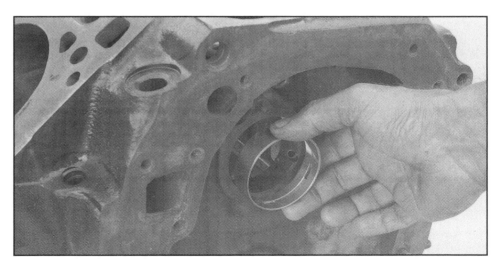
This front cam bearing has two holes and a groove in between. One is an oil supply hole and the other is for supplying oil to distributor shaft bore. Always check bearings as shown before installing them.

Driving in new cam bearings isn't much different than driving out the old ones. Just make sure you don't damage bearings as you drive them in, line up oil holes and be careful that they go straight in. Install bearings starting from front or back to middle of block with access through back or front then go to the other end to complete cam bearing installation. Cone centers drive bar in bearing bore.

Some front cam bearings must be driven in a specific distance past the front face of block. Others just have to be driven in below front face. Check engine specs. If a specification is given for front cam bearing location, use feeler gauges and a straight edge like this with thrust plate to check bearing position.

Check that cam-bearing oil holes align with those in block. If they don't, cam bearing won't receive any lubrication.

Prepare the Block & Bearings

Position the block so you have access to the cam bearing bores. With a V-type engine, this is usually with the block on its back. An inline block should be on its side.

Organize the bearings in their relative positions from front to rear. Check your notes for anything special such as oil hole or groove positioning. You'll also need some motor oil, the bearing installation tool and, possibly, a roll of masking tape. Oil holes in the inserts are sometimes slotted to accommodate misalignment with the holes in the block.

When installing each cam bearing, work from the end of the block that's opposite from the end you're installing the bearing in. That is, when installing the rear cam bearing, work through the front of the block and vice versa. When installing the center bearing—if there is one—work from either end. This gives better alignment of the drive or pull bar and, consequently, the bearing should go square into its bore. If a centering cone is part of the installation tool, great. Use it to center the bar in the front or rear cam-bearing bore. If you don't have an alignment cone, sight through the bearing bores to center the bar.

Select the front or rear bearing and the appropriate mandrel. The mandrel shouldn't fit tight in the bearing but shouldn't be loose, either. If the fit is close but too loose, take up excess clearance by wrapping some masking tape around the mandrel. When using an expansion mandrel, expand it until bearing fit is snug, then back it off slightly, or about one turn.

Check for chamfers on the outside leading edges of the bearing. If only one end is chamfered, it should go in first to provide a lead-in. Liberally oil the mandrel, put the bearing on the mandrel and align the bearing and its oil hole(s) with the bore. Make sure the bearing back and bore are clean and oil free. Center the drive bar in the bearing bore, tap on it to start the bearing, then check to see if it's going in straight. If you're using a pull bar, also check the bearing just as it enters its bore. If the bearing starts crooked, remove it and try again.

If the bearing is started straight, drive or pull it in until it looks centered in its bore and the oil holes in the bearing and block align. You may have to be more precise with positioning the front bearing. As an example, some Ford V8 cam bearings install so there is 0.005—0.020-in./0.1—0.5mm clearance between the front edge of the bearing and thrust plate. Check specs in your manual for clearances.

When you finish installing the center cam bearing, go to the opposite end of the block and work from that end to the center until all cam bearings are in place. Be careful when doing this because you must be careful not to damage the new bearings with the drive or pull bar.

Bearing Check—With all of the bearings in place, trial-fit the

Bluing and scraping cam bearings is what you'll have to do if cam is tight in the new bearings. I'm doing front one here. Center bearings are harder to do because of accessibility. This job is for an expert.

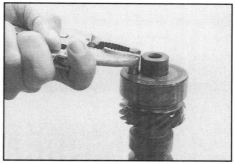

New camshafts sometimes don't come with a drive pin. If this is the case, remove pin from your old camshaft before you turn it in as a core. Remove pin by clamping on it with Vise Grips and work it out.

camshaft. Be very careful that you don't damage the bearings with the cam lobes when installing the cam.

Thread a long bolt into a sprocket bolt hole or temporarily install the sprocket so you'll have better control of the cam during this trial installation. The cam can be difficult to handle once you get it most of the way into the block. Oil the bearings and bearing journals. Use one hand on the bolt or sprocket and the other to support the cam down in the block as you move it from one bearing web to another.

When the camshaft is all the way "home" loosely install the cam sprocket if it's not already there and rotate the cam by hand. If it slid into place easily, the cam should rotate without much more than slight resistance. If the camshaft binds, however, remove it and wipe the bearings clean with a clean paper towel soaked in lacquer thinner. It's now time to perform the bluing-and-scraping process.

To scrape the bearings you'll need a tool like the one described earlier for chamfering bearings. You'll also need some machinist's blue such as Dykem. Start the process by bluing the bearings—coat them with machinist's blue. Carefully install the camshaft without oil on the journals. After rotating the cam at least one revolution, remove it.

Inspect the bearings. You'll need a small mechanic's mirror to view the full ID of each bearing. Areas where clearance is tight will be wiped clean of bluing. Such areas must be relieved. To do this carefully, scrape away a small amount of material the full length of the bearing in those areas only with a bearing scraper. Don't remove too much. Reapply bluing and check again for tight areas with the cam. Repeat the process until the cam rotates freely and without removing bluing. It should rotate without binding when you've achieved the needed clearance.

INSTALL CAMSHAFT

If you're installing a new or reground camshaft, make sure the drive pin or key is in place. If not, remove the pin or key from the old cam. Remove the pin with Vise Grips or clamp the pin in a vise and pull the cam off of it. Don't worry about burring the pin surface. Put this end in the new cam to help hold it in place. If a Woodruff key is used, tap the end of it with a small punch and hammer. It will roll out of the groove. File off any

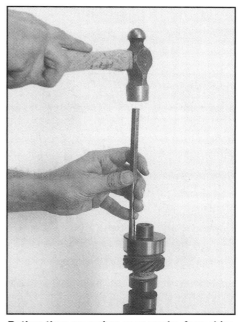

Rather than removing scar marks from drive pin, take advantage of them. Install scarred end of pin first. Drive it in with a punch and hammer until it bottoms.

burrs and tap it into the groove in the new cam.

Note: If your cam is gear driven and uses a thrust plate for end-play control, install the thrust plate as described below before going any further.

Camshaft lobes are Parkerized after being ground or reground. This chemical treatment applies a manganese phosphate solution to the lobe surfaces to give them the needed wear qualities. It hardens the lobe surfaces and increases oil retention.

Immediately before you install camshaft, coat cam lobes with moly lube. Oil bearings and bearing journals.

Unfortunately, they also retain dirt, so wipe down the cam with a paper towel and lacquer thinner before installing it.

Lubricate the lobes and bearing journals. Apply moly to the lobes and assembly lube or motor oil to the journals. Aftermarket cam manufacturers usually supply moly with their cams because they've seen first hand the sad results of improper cam lobe lubrication and break-in. So if you don't have any molybdenum disulfide, get some. Don't rely on oil to prelube the lobes and lifters.

Camshaft End Play

This is not a consideration with some engines. Two "for instances" are the big- and small-block Chevys when flat-tappet cams are used. In each of these, oil pump and lifter loads on the camshaft force the cam to the rear of the block. This holds the cam drive sprocket against the front face of the block. But change these load-producing components such as the lifters or oil pump and you'll have to install a button on the nose of the cam for thrust control.

If your cam uses a pressed-on gear, the gear must be installed with the thrust plate before the cam is installed. This is done in an arbor press, so gather up the cam, sprocket and thrust plate and head for the machine shop. Fit the thrust plate to the cam, then align the gear with the key and press it on so there's enough end play. End play should be minimal, or about 0.001–0.005 in./0.02–0.13mm. Check clearance with feeler gauges between the thrust plate and gear.

Chain Driven—For a chain driven cam using a thrust plate for end play control—it may use a thrust button—install the thrust plate now. Chains and sprockets are sometimes lubricated by oil routed through a hollow thrust plate bolt or a groove in the thrust plate. Refer to your notes concerning this.

Lubricate the back of the thrust plate and thread in the bolts. Torque the bolts to spec. Next, do a trial installation of the drive sprocket. A spacer may go behind it and a washer

If a thrust plate is used to control camshaft end play, install it and sprocket so you can check end play. A 0.009-in. (0.23mm) feeler gauge is used here.

and/or fuel pump eccentric in front of it. Again, check your notes. To torque the sprocket bolt and keep the cam from turning, insert a screwdriver through one of the holes in the sprocket web and a hole or against a web in the block. Limit end play to 0.001–0.008 in./0.02–0.17mm.

If end play is excessive, a new thrust plate should correct the

Install cam by feeding it though one bearing at a time. Be careful not to bump bearings with lobes. Sprocket is loosely installed to cam to make it easier to control cam. A long bolt threaded into nose of cam works well too.

Cam thrust plates must sometimes be oriented in a specific direction and positioned to front face of block as is the case here. It routes oil from front cam bearing to distributor-shaft oil gallery.

Once cam is in place, install thrust plate if engine is so equipped. Be careful to orient thrust plate correctly. Torque bolts to specification.

problem. Don't assume so, though. Go through the same checking procedure with it to be sure.

If your cam has a pressed-on gear, access to the thrust plate bolts is through holes in the gear. Align a hole in the gear with one in the block and install a bolt. Finish installing the other bolt(s) by doing the same.

For camshafts using thrust buttons, end play control is accomplished with the head of the gear or sprocket retaining bolt and a plate attached to the backside of the front cover. Install the cam drive gear or sprocket, any spacers or possibly a distributor drive gear and special thrust button bolt to the nose of the cam. A dial indicator at the back cam bearing journal can be used to measure fore and aft movement of the cam. To check cam end play, loosely install the front cover with gasket to the block. Check end play and gradually tighten the cover. If it's too tight, carefully grind off the head of the bolt a few thousandths at a time until you achieve the correct end play and the cover bolts are tight. If end play is excess, purchase a new bolt and start over. Remove the front cover when you've achieved the correct end play.

Aftermarket high-performance suppliers offer nylon thrust buttons or even needle bearings for selected engines. Either of these modifications are worthy of consideration if you're building a racing engine. Otherwise, save your money or put it to use elsewhere.

Note: Keep the block covered when you're not working on it. I've said this earlier, but it's even more important now that the camshaft is installed and the lobes lubricated. Protect your investment.

CRANKSHAFT INSTALLATION

It's time to prepare to install the crank, main bearings, and the rear main oil seal. What bearings you use depends on the size of the main bearing journals. The oil seal comes with the gasket set.

Size the Bearings

To determine what bearings to use, refer to your crankshaft inspection notes. There you should find measurements of the main and rod journal diameters. Compare these to specs found in the manufacturer's manual or a bearing catalog such as those from Clevite, Federal Mogul and Vandervell. There you'll find all sorts of specifications so you can determine what bearings should be used: standard or 0.010-in./0.25mm, 0.020-in./0.50mm or 0.030-in./0.75mm undersize.

Now's a good time to correct a common misconception. When a

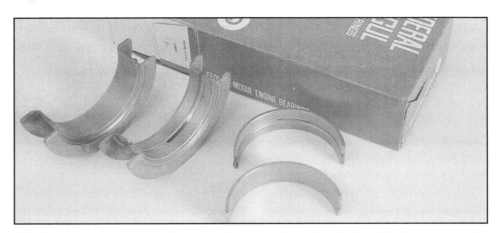

Wide flanges on one set of main bearing inserts resist crankshaft thrust loads as opposed to single-purpose radial bearing inserts at right. Holes and grooves in upper bearing halves are for lubrication.

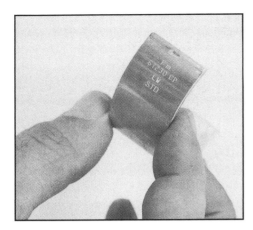

Stamped into back of bearing insert is part number and relative journal size. Bearings will be standard (STD) like this one, or undersize as indicated by -010, -020, etc.

PREPPING BEARINGS

The operation I'm about to describe makes bearing engineers weep. They prefer that a bearing insert be installed just as it comes out of the box. I agree because bearing manufacturers spend untold hours and thousands of dollars to develop their products. But most race-engine builders can't keep their hands off parts that go into their engines and bearings are no exception. And no one has proven that what I'm about to describe hurts anything. So if it'll make you feel better and you're into "romancing" engine parts, here's how to race-prep engine bearings without doing any damage.

You'll need some 3-in-1 Household Oil and newspaper. Typically fine Scotchbrite is used, but newspaper is better. Fold the paper in about a 2-inch square so it's several layers thick. Soak the paper in the 3-in-1 oil and wipe the bearing back and forth. The idea is not to remove any material, just put a bright finish on the bearing surface. To do this, use a motion with the oil-soaked paper in one hand against the bearing while it's in the other hand just as the bearing journal would if the crankshaft was rotated back and forth.

When a bright even finish is achieved, return the bearing to the box and do the same to the next bearing. Continue this until all rod and main bearings are prepped and back in their boxes.

bearing journal is ground to a smaller diameter, *undersize* bearings are used, not oversize. Remember: Use undersize bearings with undersize journals. Size is related to journal diameter, not bearing insert thickness. For example, if the journals are 0.010-in. smaller than standard, or undersize, 0.010-in. undersize bearings are used. Another point: The mains can be left standard size and the rod journals turned undersize or vice versa. But all the mains or all the rod journals must be turned the same regardless of whether they need it or not.

As a refresher, bearing journal diameters are specified in a range. An example is 2.7485–2.7495-in. where 2.7485 in. is the minimum allowable diameter and 2.7495 in. is the maximum allowable diameter. Nominal, or average, diameter of this bearing journal is 2.7490 in. If the journal size falls within the minimum and maximum range, use standard bearings. But if the journal measures 2.7480 in., it is 0.010-in. undersize, so 0.010-in. undersize bearings are used.

Two basic types of seals are used at the rear main—rope and lip types. Rope seals are two lengths of graphite impregnated rope. Similar to the main bearings, the rope seal is made up of an upper half and a lower half. One half of the seal installs in a groove in the rear main-bearing cap and the other half in a matching groove in the block. But unlike the main bearings, there is no distinction between the two halves.

Lip seals are either split or full circle. Split-lip seals are similar to rope seals in that each has an interchangeable upper half and a lower half. It too installs in a groove in the rear main bearing cap and block. Supported by a steel shell, the full-circle rear main seal installs in a counterbore in the block and rear main bearing cap or in a separate housing that bolts to the cap and block.

Sealing qualities are about the same, but lip seals are much easier to install than rope seals. Also, friction induced drag is higher with the rope seal, but this drops off considerably during the first few minutes of engine operation. One problem with lip seals is they can be installed backwards, thus resulting in a severe rear-main-seal leak.

Install Main Bearing Inserts—Mistakes in packaging and pulling parts are made. Make sure you got the right bearings. Check the back sides of the inserts. You'll find stamped on the back of the shells **STD, 010, 020** or **030** for standard, 0.010 under, 0.020 under and 0.030. To make sure the right bearings got in the box, match the numbers on the box and to those on the inserts. If you find a problem correct it before you go any further. If all is OK, get on with installing the bearings.

Position the block upside down. Using a clean paper towel soaked in solvent such as lacquer thinner, wipe the bearing bores clean. Any dirt that would prevent the bearings from fully seating must be removed. Wipe the front and back of the bearings, too. The white residue on the bearings is

OK. Just make sure both surfaces are free of dirt.

Each bearing half has a bent down tab, or lug, at one end. This tab fits into a notch at the bearing cap parting line to position the bearing in its bore, not to keep it from spinning. The top insert halves—the ones that go in the block—have oil holes and are grooved.

Install the upper bearings. Fit the tabbed end into the block first. Hold that end flush with the edge of the bearing bore with a finger or thumb, then push down on the opposite end to force the bearing into place. Check oil-hole alignment. All bearing halves will pop into place with little effort except for the thrust bearing. Its thrust flanges fit tightly over the bearing web, so additional force will be required to seat it.

Before you install the bearing inserts in the caps, lightly file the cap-to-block mating surfaces. Do this with a large flat file. Lay the file on your bench and stand a cap up on it. Hold the cap square to the file and run it back and forth while applying light pressure. The object here is to remove any nicks or burrs that could prevent the cap from seating squarely against the block, not to remove material from the cap. Break the sharp edges at each machined corner. After wiping the caps clean, install the non-grooved inserts in them. Don't install the rear main oil seal yet.

Crankshaft Prep

If you just picked up your crank after being turned or it's been in storage, give it a thorough cleaning. True, it may look great, but don't risk the chance of installing a crankshaft that is covered with abrasive grit. Wash it down with solvent and run a small bottle brush or gun-bore brush through the oil holes.

 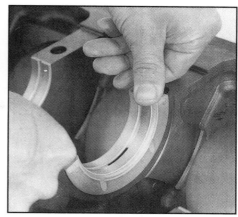

Before installing bearings, clean bearing bore in block and cap. Wipe back side of each bearing insert too and fit inserts into block and caps.

Check Bearing Oil Clearance— Now that you've determined the correct bearings for your crankshaft and have double-checked their sizes and part numbers, additional checks should be unnecessary. However, take this last chance to confirm that journal-to-bearing, or oil, clearances are correct using that little strip of wax—Plastigage. You'll need the green, or 0.001–0.003-in./0.02–0.08mm oil-clearance range.

To check clearances using Plastigage, lay the crankshaft in the block on dry bearings. That's right, no oil! Not only is Plastigage oil soluble, the oil will take up some clearance and you'll not get a true reading.

Don't turn the crankshaft once you've set it onto the bearings. Remember that the bearings are dry. Tear off a piece of paper Plastigage sleeve that corresponds to the length of the bearing journal. Remove the strip of Plastigage from the sleeve and lay it lengthwise on the journal. Don't discard the sleeve just yet. Install the bearing cap that goes in that position and torque the bolts to spec. Remove the cap to expose the mashed Plastigage. Some may stick to the bearing and some to the journal. It doesn't matter. Read clearance directly by lining up whichever graduated width at the edge of the sleeve matches the Plastigage. Read oil

 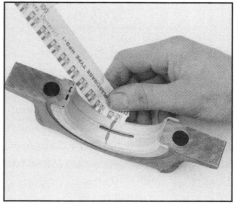

Double-check main bearing oil clearance with Plastigage. With crankshaft in place on dry bearings, lay strip of Plastigage the full length of the dry bearing journal, then install bearing cap. A very light film of oil can be applied to bearing in cap to keep Plastigage from sticking to bearing. Torque bolts to specification. Don't turn crankshaft. Carefully remove cap with bearings and check width of squeezed Plastigage with scale printed on edge of sleeve. Width corresponds directly to bearing oil clearance.

For maximum accuracy, bearings are installed in block without crankshaft, then ID is measured with dial bore gauge. Difference between bearing journal diameter and this measurement is oil clearance.

Two ways of checking crankshaft end play; with dial indicator and feeler gauges. With dial indicator, force crankshaft to rear of block using a screwdriver. Hold it there, zero the dial indicator and force crankshaft forward. Read end play directly. Using feeler gauges, hold crankshaft in one direction while you measure clearance between bearing and crankshaft thrust face.

clearance directly. The wider the Plastigage, the less the clearance.

Typically, oil clearance should be about 0.002 in./0.05mm for the average street engine, but 0.001 in./0.02mm is OK. Additional clearance, or 0.0020–0.0035-in./0.05–0.09mm clearance is desirable for a competition engine.

Crankshaft End Play—Now that you've checked main bearing radial clearances, check crankshaft end play, or front-to-back crankshaft movement. Sometimes called *float*, this movement is controlled by crankshaft and bearing thrust surfaces. End play should be within 0.005–0.010 in./0.12–0.25mm. Check to be sure.

Two checking methods are used: with feeler gauges or with a dial indicator. Whichever you use, the crankshaft and top and bottom thrust-bearing inserts need be installed. Read *Install Crankshaft,* page 122, for how to do this. The crankshaft must be in place, the thrust bearing flanges aligned, cap bolts torqued and bearings oiled—except for the thrust faces.

To use feeler gauges, force the crankshaft to the rear. Pry between a main bearing web and crank counterweight or throw and hold it there. With a 0.008-in./0.20mm feeler gauge, try inserting it between the rear thrust surfaces. If it fits with a slight drag, great. Finish installing the crankshaft. If the 0.008-in./0.20mm gauge is loose, try a 0.012-in./0.30mm gauge. All is OK if it doesn't fit. If it does with no drag, there's too much clearance. Try progressively larger gauges to find out just how much excess clearance there is. Either the distance between the crankshaft thrust surfaces is excessive or the bearing thrust flanges are too thin. Check your notes to be sure the crank is OK before going back to the bearing supplier with your complaint.

If the 0.008-in./0.20mm feeler gauge won't fit, try 0.004-in./0.10mm. If it fits with little drag, end play is OK. If it doesn't fit, you can correct the problem. But first use your feeler gauges to determine actual end play. Record it, then remove the crankshaft and thrust bearing inserts.

To use a dial indicator, mount the base to the front of the block. Position the indicator with its plunger in line with the center of the crank mains and against a flat surface at the crank nose. Using a screwdriver between a main bearing web and a crank throw or counterweight, force the crankshaft to its rearmost position and hold it, zero the dial, then force the crankshaft to its foremost position. Read end play directly. If it's within the 0.005–0.010-in./0.12–0.25mm limits you can complete the crankshaft installation. If end play is not within limits, follow the above recommendations for correcting excess end clearance. To correct insufficient end play, do the following:

To increase crankshaft end play, first determine what actual end play is and what you want to achieve. If the goal is 0.008 in./0.20mm end play, subtract actual end play from this figure and you'll know how much to thin the front thrust bearing insert flanges. This is done by sanding off material using 320-grit wet-or-dry sandpaper on a flat surface such as a piece of thick glass or a surface plate. Mike the flanges to determine what you're starting with. To determine

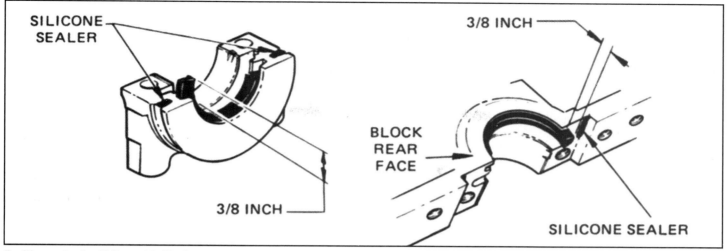

Position ends of split lip seal in block so one end is above parting face approximately 3/8 in. Half in cap should be positioned so ends will mate. Seal lips must point toward oil it's sealing. Do not overuse sealer. Apply it as shown.

final flange thickness, subtract initial flange thickness by how much you want to increase end play.

Hold the inserts end to end and move them back and forth on the sandpaper in a zigzag motion while applying light pressure. Wipe the bearings clean and measure the flanges as you progress to make sure you are removing material evenly and that you don't remove too much material. Once you've arrived at the desired flange thickness, reinstall the bearings and crankshaft after you've thoroughly washed the bearing inserts. Check crankshaft end play to confirm it is now correct.

Rear Main Bearing Seal—Except for the full-circle lip seal, you'll have to remove the crank to install the rear main seal. Regardless, the crank will have to come out so you can lubricate the bearings and bearing journals. As for how to install the seal, the split-lip type and rope seal go in before the crankshaft. I'll cover the full-circle seal later. It installs after the crankshaft.

A *split lip seal* is installed with one half in the block and the other in the cap. It must be installed so the lip points toward the oil it seals, or toward the front face of the block. Install it backwards and you'll have a major oil leak.

Note: To prevent the seal from spinning in its groove, some engine builders prick-punch the bottom of the seal groove in the block or apply a light film of RTV sealer to the back side of the seal before installing it. Use a small center punch to make two or three small prick-punch marks at the top of the seal groove.

Caution: If your engine originally had a rope-type rear main seal, check the groove in the cap for a sharp pin. This pin must be removed or flattened prior to installing a lip seal. Remove it by driving the pin through from the backside of the cap with a pin punch or flatten it using a larger punch. Support the cap on its back side while doing either. If you removed the locating pin, seal the hole with RTV silicone sealer.

A sharp edge is formed where the seal groove meets the cap parting line at each end in both the block and cap. Many gasket manufacturers supply a thin plastic "shoehorn" that prevents the backside of a seal half from being

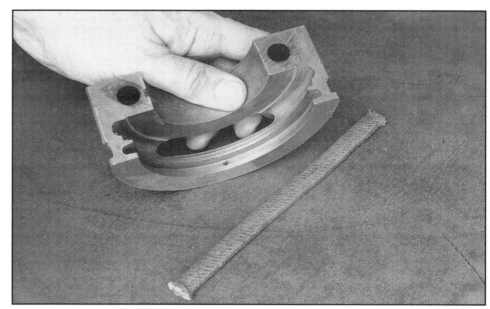

Pointed pin in rear main cap prevents rope seal from spinning in groove. If you will be installing a split lip seal, drive out pin from the back side of the cap using a 3/32-in. or smaller punch. Fill hole with silicone sealer to prevent oil from leaking out hole.

Rear main caps used with deep-skirted blocks must be sealed at each side. Slide side seals into grooves as described by instructions in gasket set.

Rear main-cap seals are compressed in grooves by installing expander, a flat strip of metal with a chamfered end. Start expander in groove behind seal with sharp edge away from seal. Push expander into groove until it is flush with block. Note full-circle rear-main lip seal.

shaved off by this edge as the seal is pushed into its groove. To use it, position the shoehorn so its tongue extends over the edge while you push the seal into place.

Unlike bearing inserts, install the seal so one end is down in its groove below the cap parting surface about 3/8 in.—the other end should project above the parting surface the same amount. Install the other seal half in a similar fashion so the seal ends will mate when the bearing cap is installed.

A *rope seal* is a bit more difficult to install. Here, the seal must be compressed into its groove. To do this form one section of rope into an half circle and lay it edgeways in the cap groove. Leave both ends projecting above the cap parting surface. There are special tools used to install rope seals, but a large diameter socket or piece of round stock and a plastic, lead or brass mallet will work as well. Support the backside of the cap and work around the inside of the seal. Roll the round section tapping as you go to gradually work the seal into place.

Check the seal by test-fitting the cap to the crankshaft. The cap shouldn't go all the way on, but close. Torquing the main cap bolts will do the rest. When you are satisfied the seal is sufficiently compressed, trim the seal ends flush with the cap parting surface using a sharp knife. Install the other section of rope in the block.

Install Crankshaft—Oil the bearings and rear main seal. Spread oil over the bearing and seal surfaces with your fingers. Give the bearing journals the same treatment, then carefully lower the crankshaft straight down onto the bearings.

If the camshaft is gear driven and the gear is pressed onto the crankshaft, align the timing marks on the cam and crank gear. Rotate the cam to where you think its mark should be, then fit the crankshaft gear to it as you lower the crank into

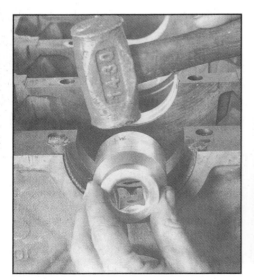

Brass hammer and large diameter cylinder like this socket are used to force rope seal into groove. Seal for cap was first installed in block to shape it, then removed and installed in cap. Work seal into groove, then trim ends flush with cap or block with a sharp knife or razor blade.

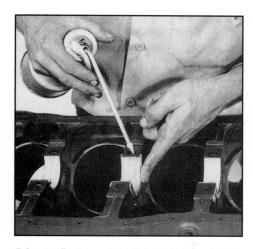

Prior to final crankshaft installation, lightly wipe off main bearings and crank journals with a clean paper towel, then coat them with oil. Lubricate rear main seal, too.

With a good grip on both ends of crankshaft, carefully lower it straight down into block and onto bearings.

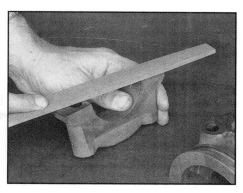

Using a light stroke, run file across bearing cap parting face. This will remove any irregularities that may prevent cap from seating fully.

position. If the two marks don't align, lift the crankshaft out, make adjustments as necessary to the cam or crank position and recheck timing-mark alignment.

Place the main caps in order. Verify that they are in the correct position and relationship using the numbers and arrows stamped or cast into each. Oil the bearings as you did those in the block and install them on their bearing journals. The caps are positively positioned in registers or on dowel pins.

Don't force the caps into position by tightening the main cap bolts. This can damage the caps and registers or dowels. If registers are used, hold one edge of the cap against its register while you strike the far corner of the cap at a 45° angle with a plastic or lead mallet. The cap should snap into place. If dowel pins or studs are used rather than bolts, position the cap on the dowels and tap the cap down into place.

Install the main cap bolts. Lubricate the threads and undersides of the heads with extreme pressure lube and drop the bolts into place. Except for the thrust bearing cap, snug the bolts then torque them to specification. Tighten the bolts in three evenly divided increments and in a pattern working out from the center cap in a back-and-forth pattern. For instance, if final torque of a two-bolt cap is 100 ft-lbs, tighten all caps to 35 ft-lbs., then 70 ft-lbs. and finally 100 ft-lbs. If four-bolt caps are used, torque the outer bolts in a similar manner.

The rear main cap should be sealed to the block to prevent an oil leak. To do this, apply a light film of RTV sealer in both corners of the register and in a line across to the ends of the seal, but stopping just short of it. Don't get sealer on the seal ends. For a block that has *deep skirts*—the block extends below the main-bearing center line—seal the sides of the cap

With bearings in place and oiled, install main bearing caps. A tap on side will seat cap in register. Don't force cap into place by tightening bearing cap bolts. You may damage block or cap. Double-check that caps are in the correct position and direction.

Be sure to tighten main cap bolts to the correct specifications, both torque and, in some cases, sequence. If four-bolt main caps are used, outboard bolts are usually torqued to a lesser amount. With cross-bolted caps, spacers between block and caps fit with zero clearance.

Torque cap with thrust bearing after seating bearing. Do this by forcing crankshaft back and forth with screwdrivers or pry bars between main-bearing webs and crank throws. Hold crankshaft forward, then tighten bolts.

after it's installed. Do this by installing a straight seal in each groove at the sides of the cap with the sealing lips toward the block. Once in the grooves, push a metal seal expander down behind each seal.

Seat Thrust Bearing—To ensure the thrust bearing flanges of the top and bottom insert align and seat, force the crankshaft back and forth in the block by prying with large screwdrivers between main bearing webs and a counterweight or throw. Alternate from side to side to force the crank forward then backward. A good rap on the flywheel flange with a heavy soft mallet while forcing the crank forward with a bar or large screwdriver is best. Don't let up. Hold the crankshaft forward in the block and tighten the thrust-bearing main-cap bolts. Torque them in three equal steps. While you have the opportunity, re-torque all main cap bolts to spec.

Unless you've installed a rope seal, the crankshaft should spin freely without binding. If it doesn't, something may be wrong, but maybe not. If you installed a rope seal, it should take about 15 ft-lbs. to turn the crank. A little more will be required to break it loose. Drag will decrease considerably after the first few minutes of initial running. But rotate the crankshaft a few turns anyway to be sure there's no binding and correct any problem before you proceed.

PISTON RINGS

Piston rings are manufactured to close tolerances for specific bore sizes, but check end gaps to make sure they are right. Not only can the wrong rings find their way into the right boxes, gaps may be inaccurate. And if you want to custom-fit end gap, you'll need to start with 0.005-in./0.13mm or 0.010-in./0.25mm oversize rings and file the ends to get the desired gap. So if bore size is 4.030 in./102.36mm, use a ring set for a 4.035-in./102.49mm or 4.040-in./102.52mm bore.

Checking End Gap

Typically, compression ring end gaps for street engines are in the 0.010—0.020-in./0.25—0.50mm range, a pretty wide margin. Exactly what the gap should be depends on bore size; the larger the bore, the larger the gap and vice versa. As a general rule there should be 0.003 to 0.004 in. end gap for each inch of bore diameter. For an engine that "builds" a lot of heat in the combustion chambers such as a stock car or turbocharged engine, end gap can be as high as 0.005 in./in. of bore diameter. So a ring installed in a 4-in. bore should have a gap of at least 3.00 in. x 0.004 in./in. = 0.012 in. With this in mind, use wider end gaps if the engine is for heavy-duty or high-performance applications. Because compression rings in the second groove run cooler, they can be gapped 0.003-in./0.08mm less. Allowable gap

Prior to checking ring end gap, push compression ring part way down cylinder and square it in bore. Push ring to bottom of its travel limit if bore is tapered. Allen Johnson uses a fixture machined from aluminum to square up ring in new bore. An old flat-top piston turned upside works as well.

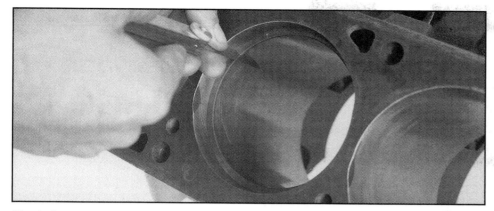

Check ring end gap with feeler gauges. A little too much ring gap is not a problem, but too little can be disastrous. If ring ends butt because of inadequate end gap, cylinder scoring will likely result.

for oil-ring side rails is much wider at 0.015—0.055 in./0.38—1.40mm.

Check ring end-gap clearance with the ring in its bore. Carefully fit the ring into the bore, then square it up with the top of an old piston or tin can. There's no need to push the ring down the bore more than a half inch unless the cylinders haven't been rebored. In such a case, push the ring down the bore about 4 inches to check end gap. This is necessary because bore taper causes ring end gap to decrease to a minimum at the bottom of its travel. A gap that appeared to be okay at TDC may cause the ring ends to butt at BDC when at maximum operating temperature. Bore scuffing or worse may result. If any gaps are too large, return the ring set for replacement. If too small and the rings are not chrome, adjust end gaps by filing the ring ends.

Filing—Always apply the filing or grinding motion from the outside edge of the ring to the inside. This is very important. If the file or grinder is moved in the wrong direction the chrome, moly or ceramic filing will chip or peel away. When filing with a straight file, don't hold the file in your hands. You won't have the control necessary to make a square end. Instead, clamp the file in a vise or onto a bench with a C-clamp or Vise Grips and move the ring against the file. When grinding, use a thin grinding disc, one that fits between the ring ends. Just make certain the disc rotation grinds the end from the outside in. To make this job much easier, special rotary files are available from K-D and Sealed Power.

Gap the second rings first, then come back to the top rings. When filing or grinding, hold the ring so you'll have good control so the ring end is cut square. Periodically pinch the ends together and hold it up to a light to check squareness. Check your progress by fitting the ring in its bore. It's easier to file. This will help ensure that you end up with a gap that right on rather than too large. Go too large and you'll have to buy a new ring set.

Deburring—Once you've arrived at the desired end gap, finish by deburring the end of the ring. Use a needle file, a whetstone or 400-grit sandpaper. The inside edge of the ring will have the biggest burr, the outside edge the least—if any—and the sides will be in between. If you can feel it, there will be a burr there. The idea here is to remove it, nothing more. And file away from the top, bottom, back and front surfaces.

After checking and adjusting end gaps—top compression ring, second

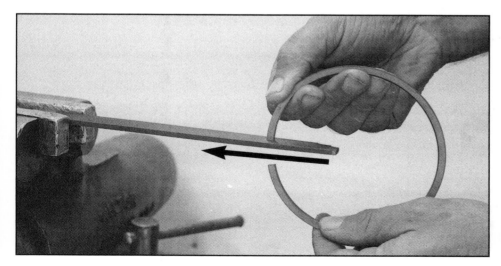
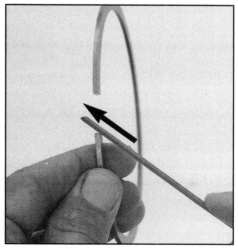

You don't need a fancy tool to increase ring end gap. Clamp fine-tooth file in vise with tang pointing away from you, then hold end of ring squarely against file and move it in direction of arrow. Squeeze ends of rings together to check that you filed ring end square. With a small file or crocus cloth wrapped around a 6-inch ruler, break sharp edges of filed ring end.

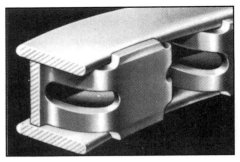
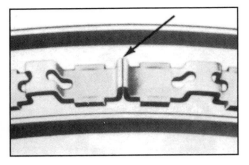

Section of oil-ring assembly. Drawing at left shows how expander/spacer fits behind rails to spring load them against cylinder wall. Ends of expander/spacer must butt, not overlap. Courtesy Sealed power Corporation.

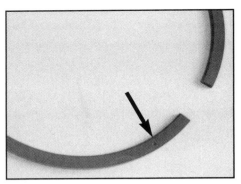

Always read directions that accompany pistons rings. Mark on ring indicates top of ring, or side that installs up.

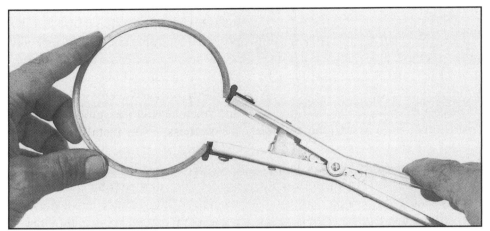

Expander makes job of spreading compression rings much easier than using your thumbs. Reduce risk of breaking ring by not spreading it anymore than necessary to get it over piston.

ring and oil rings—organize each ring set so you'll be able to install it on the piston that goes in the right bore. Repeat this process until you've checked all end gaps. Be careful not to mix up the rings. The first and second compression ring are different, but you'll have to look closely to tell the difference.

A good way to organize the rings is to drive nails in a piece of plywood in the same order and number as the cylinders in your engine. Number the nails to correspond to the cylinders and hang the rings on the nails as you fit them. If the engine is a V8, you'll have two rows of four from top to bottom. The board for such an engine should be about 20 inches long and 10 inches wide.

Install Rings on Pistons

It's possible to install piston rings without a special tool, but I recommend that you use a ring expander. Not only will it help you avoid breaking a ring and bloodying your thumbs, an expander lessens the chance of damaging a piston during installation. One thing for sure: A ring expander is cheaper than a set of rings—break one and you buy a complete new set.

To keep the piston steady while you install the rings, lightly clamp the rod in a vise between two small blocks of wood. Position the piston so the bottom edges of the skirts rest against the vise, preventing it from rocking back and forth. Be careful not to clamp on the piston. Your hands will then be free to concentrate on installing the rings, not controlling the rod and piston. If you don't have a vise, sit in a chair and clamp the rod and piston between your knees. Installing the rings in this manner is about as easy as using a vise. Following is the order of ring installation.

Oil Rings—The first ring to go on a piston is the oil ring. Usually consisting of three pieces, the oil ring is made up of the expander spacer and two steel chrome faced rails.

A ring expander isn't needed to install the oil ring rails. Save yourself some time by positioning the end gaps as you install the rings. All you need to do is find the front of the piston and determine where the ring end gaps should go in relation to it. The front will be indicated by a notch or dot in a cast piston or an arrow or the word **FRONT** stamped in the top of a forged piston.

It seems that recommended end-gap placement is different for every manufacturer. But one thing they all agree on is none of the end gaps should line up one above the other. Otherwise there would be a straight leak path which would result in increased blowby and oil consumption. Example: One manufacturer will place the top compression ring end gap over one end of the wrist pin and the second ring end gap over the other end of the pin. The oil ring rails would then be placed 90° from each other on one side of the wrist pin and the expander

Install oil rings. Expander/spacer goes on first, then top rail followed by bottom rail. Check that expander/spacer ends butt rather than overlap. Don't spiral on compression rings. And rather than using your thumbs to expand them, use an expander. Place ring ends over groove, then rotate back of ring down over piston over and into groove.

spacer ends on the opposite side. Other manufacturers suggest different end gap placement. Just use the end gap placement suggested by the engine or ring manufacturer and all will be OK.

Back to installing the oil ring. First on is the expander spacer. Note that the ends may be painted different colors or have different colored plastic blocks bonded to them. This is to make it easier to detect overlapping expander spacer ends, but be careful. They shouldn't overlap. You really have to work at preventing this from happening with some expander spacer designs.

After you have the expander spacer on, install the rails. The top one goes on first in most cases. The exception is oil rings that install in 3mm grooves. The bottom rail goes on first. Otherwise, insert one end of the rail in the oil ring groove on top of the expander spacer. Hold the free end of it with one hand while running the thumb of the other hand around the rail and spiral it into the groove. Don't let the free end of the ring scratch the piston as you bring it down into its groove. Position the end gap by sliding the rail around in the groove.

Install the second rail below the expander spacer in the same manner. Position its gap and double check that the expander-spacer ends aren't overlapping.

Compression Rings—Compression rings are less complex than are oil rings, but more difficult to install. They are so stiff it is difficult to keep from damaging them and the piston. And keep your eyes open when installing these guys. Look for identifiers that indicate the top of the ring. *Pip marks*—round or circular indentations—or a chamfer on one corner indicates the top of the ring. Read the manufacturer's instructions. Install a ring upside down and the *twist* and/or face will be backwards, causing high oil consumption, excessive blowby and accelerated ring groove wear.

Twist is the angle a ring has relative to the groove in which it sits. Twist is used for sealing combustion chamber pressures and oil control. Compression rings that have no twist have no pip marks. If you get such a set, install them either way. But inspect the rings to be absolutely sure. Also, follow the instructions accompanying the rings exactly.

To repeat, the difficult thing about installing a compression ring is spreading it so it'll fit over the piston without breaking, twisting it or gouging the piston with its ends. This is why I recommend you use a ring expander.

With or without an expander, install the second compression ring first, no exception. Make sure the pip mark is up. And don't spiral a compression into its groove as you did the oil-ring rails. To avoid breaking it, expand the ring no more than necessary and don't allow it to twist. Lower the open end of the ring over the second groove, then rotate the ring over the piston and release it into its groove. Repeat the process when installing the top ring and position the end gaps according to the ring or engine manufacturer's recommendations. Install the rings on the other pistons, position their gaps and you'll be ready to install the rod and pistons in their bores.

PISTON & ROD INSTALLATION

To install the piston-and-rod assemblies you'll need a ring compressor, two sleeves for the rod bolts, a squirt can full of oil, a tomato-like can large enough for a piston to fit in and some motor oil. Fill the can until it's about three inches deep with motor oil, or deep enough that the rings and wrist pin will be immersed when the piston is set upside down in the can. Cover the can when it's not in use.

Preparation—It's easy to get a piston and rod backwards or in the wrong "hole," so get the tools and parts organized. All tools must be within reach, the engine must be positioned just right and everything must be clean—including you. In synchronized motion, you need to

All engine components are laid out in order ready for installation on counter in race shop engine assembly room. You too can be as clean and well organized.

insert each piston and rod assembly into its bore, square the ring compressor to the deck surface and guide the rod on to its bearing journal while you tap the piston into its bore. You'll soon find it would be nice to have another set of hands.

Before you begin installing the rods and pistons, wipe out the cylinder bores. Don't use a rag. Use paper towels. Other than being soaked with oil, the towels must come out perfectly clean. Check the rods and pistons for dirt, too.

If you are assembling your engine on a bench, position the block so the deck surface is toward the edge of the bench. Block it up with 2x4s if you need to improve access to the bottom end. If it's on an engine stand, rotate the block so the crankcase is down and the deck surface level.

Organize the pistons so they are in order and within easy reach. This will prevent the inevitable search for the rod and piston that goes in a particular bore. Have the bearings and tools ready, too: ring compressor, rod bolt sleeves, can of oil and squirt can. Once everything is ready, it's time to slip the pistons and rods into their bores.

Install Rod Bearings

You can fit the rod bearings immediately before you install each rod and piston or install them all at once. It makes no difference. Just keep things clean and organized.

As you did with the crankshaft, double-check the size and part numbers of the rod bearings. Remove the rod cap and wipe out the bearing bore. Don't miss the cap parting surface. A particle of dirt here can prevent the bearing cap from seating fully, resulting in improper bearing crush. Give the connecting rod bearings the same cleaning treatment you gave the main bearings.

Place the bearing inserts in the rod and cap portions of the connecting rod. If there is an oil hole in the rod make sure the hole in the insert matches. Position the cap and nuts or bolts to their rod to ready the assembly for installation.

Position Crankshaft

Install the damper or pulley bolt and washer in the nose of the crank and leave a box-end wrench hanging from it. This will make it easier to turn the crankshaft. You need to turn it at least one full revolution to make sure nothing binds and to position the crank for the next rod and piston. It is much easier to guide a rod onto the bearing journal and install the bearing cap with the crank throw at BDC.

Position the crank and double-check the cylinder number and number on the rod and cap. Be careful not to knock the bearing insert loose from their bores. Slip the rod bolt sleeves over the bolts to protect the bore and bearing journal from the bolt threads. They'll also help guide the rod over

After wiping off bearing inserts and bearing bore in rod and cap with paper towel, install bearing inserts. If there is an oil hole in top bearing insert, check that it aligns with hole in the rod.

To pre-oil piston, immerse it in can so oil is over rings and piston pin. Rock connecting rod back and forth to work oil between pin and bore. Oil the rod bearing inserts and install rod bolt sleeves.

the bearing journal.

Liberally oil the piston rings, skirts, piston pin and bearings. Here's where the can of oil comes in handy. Hold the rod, letting the assembly hang upside down, and dunk the piston in the oil so the rings and pin are immersed. Work oil into the pin bore by rotating the rod back and forth. Wipe oil up on the skirt and the bearing inserts with your fingertips. Squirt oil in the cylinder bore and smear it around, too.

Compress Rings

Ring compressors vary in design and how they should be used, so follow the directions that came with them. If you don't have directions read on for how to use them.

Types—There are three types of ring compressors. The most common and least expensive is the ratcheting band type. This band and rotating ratchet assembly is tightened around the piston by turning the ratchet with an Allen wrench. It is released with a lever at the side of the band. The next most common type uses a thicker band with U-shaped lugs at each end. Separate ratcheting pliers engage the lugs to tighten the band around the piston and hold it in the clamped position. A lever releases the pliers.

Both compressors have a range of piston diameters they can be used with. They must also be placed on the piston correctly. The clamping device—integral ratchet or separate pliers—must be closest to the block to ensure the rings are fully compressed as they enter the bore. Additionally, the bottom of the compressor must be held square against the deck surface to keep the rings from popping out from between the compressor before they enter the bore. Oil-ring rails are the most difficult in this respect.

The third ring compressor has a cone-shaped bore. The cone tapers to a diameter that's the same as the cylinder bore. As the piston is pushed through the cone, the taper gradually compresses the rings to where they can enter the bore.

Install Pistons & Rods

Use the notch or arrow on the top of the piston and rod number and/or oil squirt hole to orient the assembly correctly in its bore. The notch or arrow must point to the front of the block; the rod number will be toward the outside of the block on a V-type engine. On an inline engine, consult your notes as to connecting rod orientation. As for the squirt hole, it should point to the thrust side of the bore, or to the right side if the engine rotates clockwise as viewed from the front and vice versa for a counterclockwise-rotating engine.

When using a band-type ring compressor, slip protective sleeves over the rod bolts and insert the connecting rod and skirt portion of the

With block positioned so bore is vertical, crankpin at BDC and rings compressed, insert rod and piston into bore. Check orientation of piston and rod—piston arrow, notch or whatever to front and rod number to the correct side. With compressor hard against block deck, push or lightly tap piston into bore with hammer handle. If you feel any resistance, STOP! Pull piston out and start over by recompressing rings.

piston into the bore. Don't let the rod bang against the bore. The oil ring should be about 1/2-inch up from the deck surface. Position the band over the piston and compress the rings. Tap the edge of the compressor sleeve down so it bottoms hard against the block deck all the way around. Using the butt end of a hammer handle, tap the piston into the bore while guiding the rod bolts over the bearing journal.

If the piston hangs up as it's entering the bore, STOP! Don't force the piston. Chances are a ring has popped out from under the compressor and is hanging up on the top edge of the bore. If you force the piston, you'll probably break a ring and possibly bend or break a ring land. Even if you're in doubt, pull the piston out and start over. All you'll lose is a few minutes rather than a ring or piston.

When the top ring enters the bore,

129

Once top ring is in bore, set aside ring compressor. As you push piston down bore, guide connecting rod over crankpin with free hand. Piston will stop with a solid thud as rod engages bearing journal. Remove rod-bolt sleeves.

Install rod cap and nuts now, but don't torque them until all rod-and-piston assemblies are installed. Rotate crankshaft 360° and position appropriate crankpin to accept the next rod-and-piston assembly. If crankshaft rotated without binding, install next rod and piston. If there is binding, investigate and correct problem before proceeding.

the compressor will relax. Set it aside and finish installing the piston and rod. Use one hand to tap the piston down with the hammer handle while you guide the rod square onto the bearing journal with the other hand. You'll know it when the rod is fully seated.

For a cone-style compressor, slip the piston into the cone far enough that the skirts extend out the bottom. You can then insert the rod and skirts in the bore and proceed as above.

Remove the sleeves from the rod bolts and fit the cap to the rod after oiling the bearing. With one hand, hold down on the piston and use the hammer handle with the other hand to tap on the cap to seat it. Install the nuts or bolts and snug them. Torquing will come later. Double-check that the cap and rod numbers coincide. If used, check the squirt hole or bearing offset, too. Give the crankshaft a full turn to check for binding, then position the crank so the next throw is at BDC for installing the next piston.

If you detect any binding when turning the crankshaft, don't go any further without finding and correcting the cause. More than likely, you installed the last rod or cap backwards. Don't be fooled, though. As you install more pistons and rods, it's natural that additional torque will be required to turn the crankshaft.

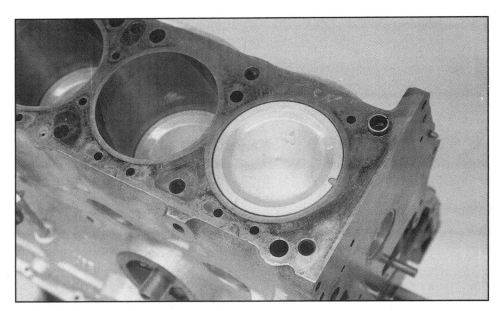

Double-check rod and cap numbers as you go. They must match and point in the right direction as should piston notches or arrows. Even the most accurate pistons or bores are not exactly the same diameter, so measuring and matching them gives consistent piston-to-bore clearance.

Don't be disturbed by this unless the turning resistance out of proportion to the number of pistons you've installed or binding occurs in one spot. Check to make sure and correct the problem.

Rod Side Clearance— This is checked after you have all pistons and rods in place. This dimension varies depending on whether an oil squirt hole is used in the rods. This hole will be drilled in the big end of the rod or as a notch in the rod or cap parting surface. If a squirt hole is used, rod side clearance should be 0.007–0.013 in./0.18–0.33mm. If not used side clearance will be 0.010–0.020 in./0.25–0.50mm. Piston lubrication depends on oil throw-off from the crankshaft, necessitating additional rod side clearance when a squirt hole is not used.

Finish rod-and-piston installation by torquing the rod nuts/bolts. Insert a feeler gauge that equals side clearance on both sides of the rod while doing this. It will keep the rod and bearing square to the bearing journal and prevent uneven bearing loading during torquing.

TIMING CHAIN & SPROCKETS

The key(s) in the nose of the crankshaft must be in place before you can install the timing set. Install it if it's not there. The same goes for the cam drive pin or key. It should be there, but if not, install them.

Two types of keys are used to drive the crankshaft sprocket: a straight key and a Woodruff key. The straight key fits in a straight keyway. The Woodruff key is half-moon shaped as is its keyway. Remove any burrs on the keyway and key with a fine-tooth file and tap them in place with a soft mallet or hammer and soft punch.

To install the timing set, the cam must be timed to the crankshaft. On

Check rod side clearance. This should be less for rods with squirt holes.

some engines this is easily done by positioning the crankshaft so the crank-sprocket key is up and the cam-sprocket drive pin is down—they point at each other. Other engines have a different relationship between the crankshaft key and camshaft pin, so this procedure won't work. But regardless of key position, the number-1 piston will be at TDC. To eliminate any doubts as to correct crank and cam relationship, do a trial installation of the crank and cam sprockets.

Check the crank sprocket. The front side will have a timing mark—a triangle, round or square indentation—on one of the teeth.

After installing all rods and pistons and checking that crankshaft rotates freely, torque rod bolts. Rather than using a torque wrench on this high-performance engine, rod bolt stretch is used as a more accurate indicator of bolt tension.

Rotate crank and cam so they are both at their TDC positions prior to installing timing set. In this case key in crankshaft points at cam and drive pin in camshaft points at crank.

With chain hanging from cam sprocket and crankshaft sprocket laying in chain so timing marks align, fit crank sprocket to crankshaft nose. Engage keyway in crank sprocket with key in crankshaft, then cam sprocket with pin or key in cam. You may need to rotate crankshaft one way or the other a slight amount to align cam sprocket.

Secure cam sprocket with bolt and washer. If your engine uses a separate fuel-pump eccentric, install it too. Note sprocket timing marks

Slide the sprocket onto the crankshaft nose and rotate it into engagement with the key. If the crank sprocket doesn't slide onto the key and crank nose, no more than light tapping with a soft mallet should move it. Stop short by about 1 inch of installing the sprocket all the way.

Note: If you're using a double roller chain, the crankshaft sprocket may have three keyways: one for standard cam timing and the others for advancing or retarding the cam 4°. I recommend that you use the advanced position for better bottom-end performance.

Rotate the crankshaft so the timing mark is up, or pointing at the cam nose—number-1 piston should be at TDC. If it's not still there, install the damper bolt and washer so you can rotate the crank. Next, loosely install the cam sprocket and find its timing mark. Here's where manufacturers differ. Some time the cam to the crankshaft at TDC so the timing marks on both sprockets point at each other; others have both marks pointing up. To dispel any doubt, install lifters on the intake and exhaust lobes for cylinder-1 and rotate the cam so both are down, or are on the lobe base circles—between compression and power strokes. If you see any lifter movement with slight cam movement, rotate the cam 180° and check again. Now loosely install the cam sprocket and check the position of the timing mark. Although this won't accurately position the cam, it will give you a general idea of the intended timing-mark positions. If the marks are not exactly straight up or down, rotate the camshaft to bring the timing marks into alignment.

Now for the chain. Remove the cam sprocket and drape the chain over it. Hold the cam sprocket so the chain is under the crank sprocket. Lift up the cam sprocket so the chain engages the crank sprocket and the timing marks align. Slide the crank sprocket back on the crankshaft nose and guide the cam sprocket on to the cam nose. Be prepared to rotate the crankshaft slightly to align the cam sprocket. Once both sprockets are in place, check the timing marks. If they align, complete the cam sprocket installation.

Some manufacturers use an oil *slinger*—a sheet-metal disc designed to throw off excess oil—on the nose of the crankshaft. If used, this disc will install in front of the crank sprocket and engage the key. The crank damper will install in front of the slinger, clamping it in place. If your engine uses an oil slinger, don't forget it. Otherwise the front seal will be flooded with oil and a leak will result.

Install and torque the cam sprocket bolt(s). If a separate fuel pump eccentric is used, it installs behind a single sprocket attaching bolt and washer. The fuel pump eccentric will engage the sprocket dowel or a tab on the eccentric will engage the dowel hole. While you have the opportunity, oil the timing chain. Give the fuel pump eccentric a good coat of oil, too.

DEGREEING A CAMSHAFT

Degreeing in a cam is a process that most automotive enthusiasts are aware of, but few actually understand. To eliminate any doubts let's look at the objective of degreeing in a cam before we look at the details of the degreeing process. To do this you must understand the specifications that describe the features of a camshaft. These are what you must verify.

Most cam specifications are listed on a cam card that is supplied with high performance aftermarket camshafts, page 137. The intake and exhaust valves open, close and reach maximum lift in direct relationship to the position of the crankshaft in degrees . . . degrees of the crankshaft, not the camshaft. You need to verify that these events occur in your engine as specified on the cam card. Other listed specifications typically include gross valve lift (zero lash with mechanical lifters), lobe lift, lifter lift between which opening and closing events occur as measured in inches and, from this, duration of intake and exhaust valve openings in degrees. Another spec is *lobe separation*, sometimes called *spread*, or the distance in degrees between the intake and exhaust valve lobes at maximum lift of each in *camshaft degrees*, not crankshaft degrees. Next is valve lash if the cam is used with solid lifters.

Rocker Arm Ratio

Another term with which you need to be familiar is rocker arm ratio. It determines the difference between lifter movement at the cam lobe and actual valve lift, which is the larger of the two. Typical rocker arm ratio is 1.5:1. What this means is the valve opens 1-1/2 times the full movement

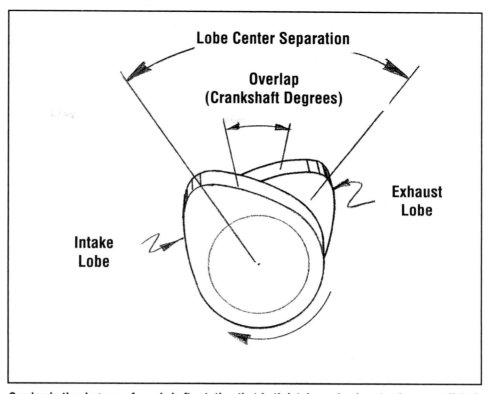

Overlap is the degrees of crankshaft rotation that both intake and exhaust valves are off their seats at the same time. Lobe separation is the amount of camshaft degrees between the intake and exhaust lobes at maximum lift.

of the lifter, less lash if it's a solid lifter. So if lobe lift is 0.300 in., maximum valve lift is 1.5 x 0.300 in. = 0.450 in. To illustrate further, if a rocker arm with a 1.75:1 ratio were substituted, maximum valve lift becomes 1.75 x 0.300 in. = 0.525 in. Sounds like you're getting something for nothing. Not quite. Valvetrain loads go up, thus friction and all the bad things that accompany it: wear and lost power. Two other factors must be considered: increased valve spring compression and reduced valve-to-piston clearance. Damaged valves or valvetrain may result if these are not allowed for.

The first thing you must do in the process of degreeing in a cam is to find the exact top dead center of cylinder number 1. Once cylinder-1 TDC is established and indicated with a degree wheel and pointer, you can turn your attention to the cam. You then use a cam follower on the intake and exhaust lobes with a dial indicator in conjunction with the degree wheel to verify timing and lift specifications listed on the cam card.

Checking a cam is ultimately up to the engine builder . . . you. And the checking must be done with the engine in which the cam is to be installed. This is simply because engines are different, too, and these differences can affect valve timing. So if there are timing errors caused by the relationship between the crankshaft and camshaft, you must be the one who finds and corrects the problem in your engine. Just in case you're wavering on whether or not to degree in your cam, think about the wasted hours you can avoid by finding out now that the cam must be removed rather than later when you discover your engine is down on power because of improperly ground or indexed lobes. Assuming you are convinced, let's get on with how to degree in a camshaft.

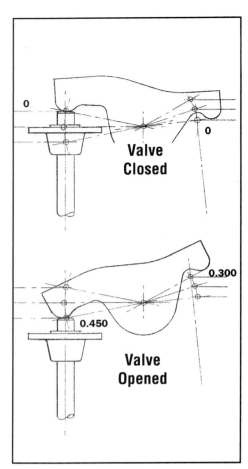

Rocker arm ratio is the distance the valve end of the rocker arm moves compared to the pushrod end. Typically this is 1.50:1. In this case, if the pushrod end moved 0.300 in. the valve tip end moved 0.450 in., or 1.5 times 0.300 in. Higher ratio rocker arms are available for some engines. For instance, if a 1.7:1 rocker arm was substituted for the 1.5:1 rocker arm, valve lift would be 0.510 in.

Degreeing Tools

As you might guess, the first thing you need for degreeing in a cam is a degree wheel. Supplied by many high performance parts manufacturers, degree wheels can be purchased from your local hot rod shop or through mail-order. In addition to the degree wheel, you'll need a 1-inch minimum travel dial indicator and a magnetic or clamp-on base to support it. This is for measuring lobe lift and, possibly, to find TDC.

Next on the tools list is a reworked lifter—not needed if the cylinder head is installed. You can make the checking lifter yourself. Remove the pushrod socket and any other lifter internals from the lifter body. In the lifter body, install a piece of steel or aluminum round stock that's slightly smaller than the lifter body bore. Insert it in the lifter body and retain it with epoxy or a derivative such as J-B Weld. It can project out of the lifter if you like but the end must be flat and square to the lifter centerline. To make sure of this, face the piece on a lathe before you install it in the lifter body.

Caution: Do not use a flat-tappet lifter to check a roller cam and vice-versa. This is OK if you're only checking lobe centerline and lift, but all other events and lifts will be inaccurate from initial lifter movement off the base circle to all events just before and after maximum lift. For a roller cam you'll need two lifters if a guide bar is used between the two. You'll only have to modify one, though. If a bolt-in guide is used such as those found in OEM engines, you'll only need one roller lifter.

Positive Stop—If you settle on using the positive-stop method of finding TDC—decide after reading the following—you'll need to fabricate an adjustable stop. Make a strap from a 1-1/4-in. wide, 1/2-in.-thick steel strap that's at least 6-in. long. Drill a hole in each end so they match the center distances of two head bolt holes and the strap spans the cylinder bore. At the center of the strap drill and tap a hole for a 1/2-13 thread. In the hole insert a 1/2-13 bolt with a jam nut on it. Grind a generous spherical radius on the end of the bolt so the piston will have a smooth surface to contact. If domed pistons are used, space the strap up from the deck surface using an oversize nut or stack of flat washers over the mounting bolts between the strap and deck surface.

To find TDC with a cylinder head installed, you'll need a stop that screws into the spark plug hole. These, like degree wheels, can be purchased, but you can make one from an old spark plug, a bolt and a nut. Just knock out the center of the plug, porcelain and all. Next, tap threads in the center of the plug the size that has a tap drill closest to the ID of the plug shell and install the appropriate bolt. As above, grind a spherical radius on the end of the bolt, run a jam nut on it and thread the bolt into the spark plug shell.

Pointer—A piece of heavy wire such as that from a coat hangar will do. An 8-in. length should be more than enough. You can determine final pointer length once you have the degree wheel in place and one end of the wire bolted to the front of the block. Use the appropriate size bolt with a flat washer in a hole for the front cover or water pump. Loop one end of the wire to fit around the bolt and grind or file the other end to a sharp point after you determine the best length for the pointer.

Finding TDC

Positive Stop Method—If they aren't in place, the crankshaft and piston-and-rod assembly must be installed. The same goes for the camshaft and timing set. The camshaft should be installed in the *straight up* position, not advanced or retarded. Otherwise, you'll have to compensate for the amount of advance or retard during the degreeing process. Read on for how to install these components. If your engine is assembled, remove all spark plugs so the crankshaft will be easier to turn.

Install the degree wheel to the crankshaft nose using the damper bolt and washer. Turn the crankshaft until you think the piston is at TDC. You can come pretty close by putting a

Allen Johnson uses positive stop on piston and degree wheel with wire pointer to find TDC on this racing engine. Method is quicker and is preferred by most high-performance engine builders to dial in.

finger on the deck surface adjacent to cylinder 1 so the tip hangs over the bore. Run the piston up until it contacts your finger. Keep turning the crank until you feel the piston start back down, then back it up slightly. This should place the piston very close to TDC. Align the pointer to TDC on the degree wheel by bending the pointer or rotating the degree wheel. Run the piston half way down the bore and install the positive stop. Remember to allow clearance for a domed or pop-up piston and space up the stop as necessary. If the cylinder head is installed, install the positive stop in the number-1 spark plug hole.

Check that the stop bolt is secured with the jam nut, then slowly turn the crankshaft clockwise until the piston contacts the bolt. Note and record the degree wheel reading. It should be so many degrees before TDC. Now turn the crank counterclockwise until the piston comes back up and contacts the stop bolt.

If you guessed correctly and got the TDC mark right on—highly unlikely unless this is not the first time you've done it—the reading should be the same number on both sides of TDC. More than likely they might be something like 22° before TDC and 26° after TDC. In this case average the two numbers by adding them and dividing by two, or 22° + 26° = 48°, then 48°/2 = 24°. With the piston still up against the stop, bend the pointer or rotate the degree wheel so the averaged number is indicated, or 24° in the example. True TDC will now be indicated. Double-check to make sure. Turn the crank back the other way against the stop and check the reading. It should also indicate the same number, or 24° in the example. If the numbers don't match, go back through the process until they do.

Once the before-and-after-TDC numbers balance, remove the stop. Rotate the crankshaft and align the TDC mark with the pointer. The piston will be on TDC. You can now degree in the cam.

Head Off, Dial Indicator Method—Although not as simple, you can use a dial indicator to find TDC in a manner similar to that used with the positive stop. Start by mounting the dial indicator so the plunger is square to the bore and its tip rests against a flat part of the piston. In this case you run the piston all the way up to TDC rather than stopping short.

Turn the crank and watch the movement of the indicator. When the piston reaches true TDC as you rotate the crank clockwise, the rotation of the indicator needle will reach a maximum, then it will slow and reverse direction as TDC is past. Note the point at which this reversal occurs and reposition the crank to put the indicator needle at this maximum position. Now, zero the indicator by rotating its dial face. At the degree

One method used to find TDC is with dial indicator on piston and a wrench to turn crankshaft. Although not yet installed, bolt degree wheel to crank nose and attach pointer to front face of block. As piston passes TDC, indicator pointer reaches maximum and swings back in other direction. Trick is to have pointer indicate 0 the instant it reverses direction.

wheel, bend the pointer or rotate the wheel so TDC is indicated. Rotate the crank almost 360° clockwise and stop 0.020 in. short of zero on the dial indicator. Note and record the degree wheel reading. It will be so many degrees before TDC. Continue turning the crankshaft clockwise so the dial indicator again reads 0.020 in. before zero and note the degree wheel reading. The degree wheel should read the same degrees, but it will be after TDC. If it doesn't, average the two numbers as above by adding them and divide by two. Reposition the degree wheel so the pointer aligns to this average number. Now when you rotate the crankshaft so the indicator reads zero and the degree wheel is at TDC at the same time, the piston should be at true TDC.

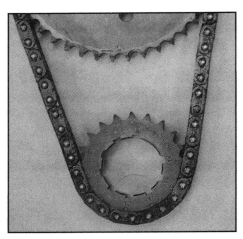

Crank sprocket's nine keyways are for timing camshaft straight up as well as four advanced and four retarded positions in 2° increments. Maximum advance and retard is 8°. Cam timing for best performance is found through a combination of dynamometer and track testing.

Degreeing in Cam

Finding Lobe Center—Because intake valve events are by far the most critical to engine performance, it's the cylinder-1 intake lobe that is normally checked. If you don't take anything for granted—like in don't assume—you may also want to check the exhaust lobe for cylinder 1. If you're a total fanatic, you'll check all the lobes for all the cylinders. That means you must also find TDC for every cylinder, too. In this case you will probably find an error if there is any before you get too far.

Begin by installing the checking lifter on the intake lobe for cylinder 1. Lubricate it with a light oil such as WD-40 or 3-in-1. If you have any doubts about which lobe is which, look at a cylinder head. If the intake valve is at the end, put the lifter in the first lifter bore. Otherwise, put it in the second lifter bore. The crankshaft should still be on TDC. If it's not, put it there. Also, the lifter should be on the base circle of the cam—it should be low in its bore. This means cylinder-1 is at the top of its firing stroke. If it's not, turn the crankshaft 360° clockwise and align the TDC mark. From now on always turn the crank in the clockwise direction, not counterclockwise, because of back lash in the timing chain or gears.

Set up the dial indicator so the plunger is square against the lifter and there's some plunger travel, but no more than 0.500 in. for a 1.000-in.-range dial indicator. If the intake manifold is installed, install a pushrod and set up your dial indicator to it. Rotate the crankshaft clockwise until the tappet is at maximum lift—or the indicator needle just begins to reverse direction. zero the dial indicator and note the degree wheel reading. It should be in the 100°–120° after TDC (ATDC) range, or agree with that listed on the cam card for when maximum lift occurs, following page.

Find the cam-lobe centerline by turning the crankshaft clockwise two full revolutions until indicator reads the manufacturer's checking clearance—0.020-in., 0.050-in., or whatever lobe lift is specified—for the first time. Note and record the degree wheel reading. Continue turning the crankshaft past the point of maximum lobe lift until the manufacturer's checking clearance is again indicated. Note and record the second degree wheel reading. The two readings should have bracketed the specified crank angle. For example, if maximum lift should have occurred at 110° ATDC the indicator readings may have been 70° ATDC and 150° ATDC, or the same amount on both sides of the crank angle where maximum lift occurs. To calculate this figure, add the two numbers and divide by two, or 70° + 150° = 220° and 220°/2 = 110°. The cam centerline is at 110° crankshaft angle.

Finding Opening & Closing Points—This method of checking cam timing is a little more involved than finding the cam-lobe centerline, but it can be use for both symmetrical and asymmetrical cams. Symmetrical cam lobes have the same *profile* on the opening and closing ramps; asymmetrical lobes do not. Using the lobe centerline method is not recommended for checking asymmetrical cams because of the different opening and closing ramp profiles.

While watching the dial indicator, rotate the crankshaft clockwise until the indicator needle is at its lowest reading—the lifter will be on the lobe base circle. Zero the indicator. Again, rotate the crank clockwise until the checking clearance is indicated, or 0.050 in. Also note and record the reading at the degree wheel. It should indicate so many degrees before TDC (BTDC), or for discussion purposes 20° BTDC. Continue turning the crank clockwise until 0.050 in. is indicated again, which will be so many degrees ATDC, or say 45° ATDC. These two numbers should match those on the cam card. If they do, cam timing is OK. Add these two numbers and 180° to get duration, or 20° + 45° + 180° = 245° duration.

Correcting Valve Timing—What do you do if the figures on the cam card and yours are different? This is a good news, bad news situation. The good news is you found a problem that exists. The bad news is the cam is out of time or the opening, closing and lobe centerline results don't match the card specs. The good news is you can probably fix cam timing—this is where all events are out of spec by the same amount and in the same direction. There are several methods of correcting cam timing. As for correcting events that don't match, you can't. But you can go back to the manufacturer and get the cam replaced. Just make sure your numbers are correct before you raise a fuss.

HOW TO READ A TIMING CARD

Every aftermarket camshaft includes a spec sheet called a *timing card*. While each company prints their cards differently, most of them detail the same information. The timing card lists specific opening and closing points for both lobes as well as other information that is usually not listed in the catalog. The timing card information is useful for both reference and to use when degreeing in the camshaft. We are using a Comp Cams timing card for this example. The numbers below correspond with those on the chart.

1. Camshaft part number and application
2. Grind Number is often different from the part number. This is usually the number stamped on the end of the camshaft.
3. Valve Adjustment will specify a lash clearance if the camshaft is a mechanical cam. In this case, the cam is a hydraulic which requires lifter preload.
4. Gross Valve Lift is determined by multiplying the lobe lift times the stock rocker ratio of 1.5:1. In this case, Comp Cams has rounded the lift off to 0.460 inch (0.3026 x 1.5 = 0.454)
5. Duration at 0.006 Tappet Lift is the duration measured between 0.006 inch of tappet lift on the opening and closing side. Compare this to Duration at 0.050.
6. Valve Timing is given on this sheet at 0.006 inch tappet lift for the intake and exhaust opening and closing points. Other cam companies such as Crane list the opening and closing points at 0.050 inch. Note also that these opening and closing points are for the cam installed with a 106° intake centerline. Moving the centerline will move the opening and closing points as well.
7. Duration at 0.050 is the duration of each cam lobe at the 0.050-inch tappet lift checking height.
8. Lobe Lift is the actual lift as measured at the lobe.
9. Lobe Separation is the angle in cam degrees between the intake and exhaust lobe. This is one way of referencing valve overlap.
10. In this case, Comp Cams also makes a recommendation for matching Comp Cams valve springs that complement this camshaft.

1—PART # 12-210-2
ENGINE: CHEV. SML BLK 265-400

2—GRIND NUMBER: CS 268H-10 HIGH ENERGY

	INTAKE	EXHAUST
3—VALVE ADJ.:	HYD.	HYD.
4—GROSS VALVE LIFT:	0.460	0.460
5—DURATION @		
0.006 Tappet Lift:	268°	268°

6— VALVE TIMING:

@ 0.006 Tappet Lift	OPEN	CLOSE
Intake:	28° BTDC	60° ABDC
Exhaust:	68° BBDC	20° ATDC

These specifications are for a cam installed at 106° intake center line

	INTAKE	EXHAUST
7—DURATION @ 0.050:	218°	218°
8—LOBE LIFT:	0.3026	0.3026
9—LOBE SEPARATION:	110°	

10—Recommended valve spring no. 981 (valve spring specs. furnished with springs.)

If cam timing is way out, check to make sure the cam sprocket/gear timing marks align. It's possible that they are a tooth off. If not you'll have to rotate the cam or crank sprocket on the cam or crankshaft, respectively. To do this, I find it easiest to use an offset crank sprocket/gear key. These keys, which are available through mail-order houses or from your local hot rod shop, come in enough degrees of offset that you'll be able to adjust the timing of your camshaft so it's right on. Another possibility is to use a crank sprocket with multiple keyways. This you may already have. Some crankshaft sprockets have up to nine keyways that allow a camshaft to be advanced or retarded in 2° increments up to a total of 8°. More common is the three-keyway sprocket. It allows a cam to be advanced or retarded 4°. Finally is offset cam bushings. They come in varying degrees of advance or retard. This requires that you drill out the dowel pin hole in the cam sprocket/gear so it will accept the bushing. Install the bushing, keyway or whatever so the cam is moved in the direction of engine rotation relative to the crankshaft and timing advances; move it in the reverse direction of engine rotation and timing retards.

VALVE-TO-PISTON CLEARANCE

Clay Method—If you are installing a high lift camshaft or oversize valves, you must check valve-to-piston clearance. An interference condition between the valves and pistons can cause severe damage to the valves, pushrods, pistons and bores. Immediately after degreeing in a cam is an excellent time to make this check since most of the necessary tools, parts and equipment are in place.

The traditional way of checking valve-to-piston clearance is by temporarily installing a cylinder head with valves, lifters and pushrods for cylinder 1. Before you install the head, apply an 1/8-in. slab of modeling clay to the top of the cylinder-1 piston. Apply flour, some cooking spray (such as PAM®) or simply smear motor oil over the clay to help keep it from sticking to the valves. Set the crankshaft on TDC of the power stroke and install the head with a new gasket. Torque the bolts to spec. Lubricate the lifter bases with moly, install them, their pushrods and rocker arms, then set valve lash to zero. Note: Use solid lifters for making this check even though they may be on a hydraulic cam.

Rotate the crank at least four revolutions, but stop if you feel any resistance. Don't force it. You may be able to see if piston-to-valve contact is the problem by looking into the spark plug hole or intake or exhaust port. If you suspect the piston contacted a valve, remove the head and check to make sure. Otherwise keep turning the crankshaft so it's back to TDC for the fourth time. The lifters should be on the base circles of the cam. You can now remove the head.

There should be an impression from each of the valves in the clay. The question is how close did the piston come to them. To determine this, section the clay at the center of each valve impression with a sharp blade such as a X-Acto knife or razor blade. The thickness of the clay will tell you the piston-to-valve clearance for each.

Check piston-to-valve clearance by measuring the thickness of the clay. This is best done with the depth gauge end of vernier calipers. Also check that the valve reliefs, or notches, are large enough in diameter, particularly if you've installed larger valves. Clearance for the intake and exhaust

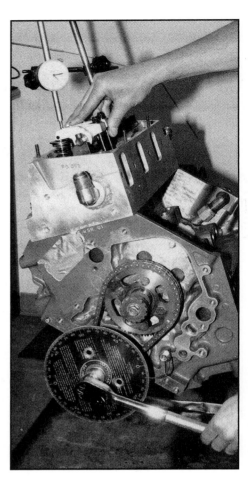

Allen Johnson uses valves installed with light springs, lifters and pushrods to check valve-to-piston clearance. Dial indicator with plunger set square against spring retainer and parallel to valve stem is used to read valve movement and thus valve clearance. At each 2° increment of crankshaft rotation between 20° before and after TDC, valve is pushed open until it touches piston. Clearance is read directly from dial indicator.

valves should be 0.080 in. and 0.100 in., respectively. The reason for the additional clearance to the exhaust valves is they are being closed by the valve springs as the piston is approaching TDC. The intake valves are just opening. Consequently, if the valves float, the chance of an exhaust valve "kissing" the piston becomes much higher than for an intake. Where both valves are at risk is if hydraulic lifters are used. If pump-up occurs, both can be held off their seats, thus valve-to-piston clearance will be reduced.

If you find clearances to be insufficient, leave the clay on the checking piston and deliver all of them to a machine shop that specializes in performance engines. This way the machinist can see the problem and enlarge or deepen the valve reliefs accordingly.

Dial Indicator Method—This is a valve clearance checking method that some engine builders prefer over the "clay mashing" method. One thing for sure is it's more precise. One thing you can't check for is clearance to the edge of a valve relief. The pistons must be cut for the size valves being used.

Preparing for this method of piston-to-valve clearance is still the same as the clay method in that you temporarily install a head, the degree wheel must be in place and you turn the crank. The difference is you must install light springs that can be compressed with your fingers in place of the standard valve springs. These are available at most hardware stores. Additionally, set up a dial indicator with the plunger square against one of the spring retainers.

Install the head, lifters and pushrods as above. Again, the crankshaft should be on TDC of the firing stroke when doing this. Adjust valve lash to zero. Install a dial indicator so the plunger tip is against a flat section of the spring retainer and the plunger is parallel to the valve stem. The plunger should be compressed slightly more than valve lift. Zero the indicator dial.

What we're going to do now is check exact valve-to-piston clearance by pushing in on the valve until it contacts the piston at the top of its intake stroke. Clearance is read directly from the dial indicator while the valve is in contact with the piston. This is done in crank position increments of 2° between 20° before and after TDC. Rotate the crankshaft clockwise, but stop 20° before TDC of the intake stroke. You'll need to record crank position and corresponding dial indicator reading for each valve. Make up a list of the 21 crank positions in a column—20°, 18°, 16°, . . . 2° BTDC, 0°, 2°, . . . 16°, 18°, 20° ATDC—and list beside of each under the relative **INTAKE** and **EXHAUST** indicator readings. If the minimum reading exceeds 0.080 in. for the intake valve and 0.100 in. for the exhaust valve, clearance is sufficient. On the other hand if an actual valve-to-piston clearance is under the minimum, machine work must be done to deepen the piston valve reliefs.

TIMING-CHAIN COVER & CRANK DAMPER

If you haven't replaced the seal in the front cover do it now. The gasket and seal surfaces should be clean and free of old gasket material. Check also that the areas around the bolt holes of a sheet-metal cover are not distorted. Flatten these areas with a hammer and flat punch.

Remove & Install Seal

Remove the front seal with a punch and hammer. In order not to bend or break the cover, support its back side by laying it over the open jaws of a vise, a short section of pipe or between two wood blocks. Check that there's enough room for the seal to pop out. Work around the seal and gradually drive it out with the punch.

You'll find a new seal in the gasket set. This full-circle lip seal must be driven in so as not to damage the seal or its outer steel shell. First, coat the seal OD with grease. This will seal its OD and ease installation. Again, support cover directly in front of or behind seal bore depending on whether seal is driven in from the back or front. Position the seal over bore so the large lip—there may be a smaller one—will point toward the crankcase when installed.

If done carefully you can tap lightly with a hammer around the outside edge of the seal and work it into place. Better yet, find a pipe, tube or large socket that has an OD slightly smaller than the seal OD and drive it into place. Don't let the seal cock as

Fuel-pump eccentric prevents front crankshaft seal from being flooded with oil. It goes on before front cover and damper. Don't forget it or front seal will leak.

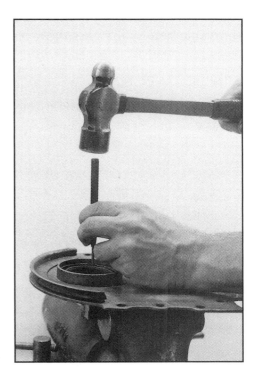

Carefully remove lip seal from front cover. Aluminum covers are easily broken and sheet-metal covers are easily bent. Use a hook-type seal remover or punch and hammer and work seal out a little at a time.

Apply grease or RTV to seal OD and position seal over bore in front cover. Seal lip should point toward engine side of cover. While cover is firmly supported on back side, drive seal in from other side. A large-diameter socket and brass hammer is used here.

Good news! If damper seal surface is badly grooved or otherwise damaged, you don't have to buy another damper. This thin metal sleeve from Fel-Pro installs over damaged seal surface.

Apply RTV to the sleeve ID and drive it over original seal surface until it bottoms. Use a wood block between sleeve and hammer. Damper will be like new.

you drive it in. It must go in square to the correct depth to ensure the seal won't leak at its outer periphery.

Dowel Pins—Some front covers are positioned on the front face of the block by dowel pins. This centers the seal on the crank nose. Others must be centered with a special tool that fits on the crank nose and into the front cover seal. If yours uses dowel pins, install the front cover now. Otherwise, have the damper ready to install with the front cover. The damper will act as the "special" tool. If this is the case with your engine, prep the damper for installation as described below and continue with the front cover installation.

Gasket Sealer—Find the front cover gasket(s) in the gasket set and check their fit to the cover. The only place sealer is needed is around water passages to seal both sides of the gasket(s). If your front cover doesn't have these passages, sealer is not needed. It will, however, come in handy. A few dabs of RTV silicone will hold the gasket(s) in place until you can get in at least four bolts. Leave these bolts loose if the seal must be centered. Fit the damper to the crankshaft as described below and tighten the front cover bolts to spec.

Prepare Damper—Two areas of the damper require your attention: the rubber bond between the damper hub and outer ring and the condition of the seal surface. Check the rubber bond for deep cracks and looseness. Surface cracks aren't a terminal problem. However, if the ring is loose on the hub, replace the damper. To check for looseness, slide the damper on the crank. Try rotating the outer ring around the hub, then try twisting the ring back and forth on the hub. If you don't detect any movement the damper should be OK.

The seal surface should be smooth and free of deep grooving. If all appears OK, restore the seal surface by smoothing it with 400-grit emery cloth or crocus cloth. This gives the surface a fine tooth that will retain oil and lubricate the seal without damaging it. If the seal surface is deeply grooved, not to worry. The damper can be salvaged by installing a thin-wall sleeve over the damper nose. Buy this sleeve at your local auto parts house.

To install the damper repair sleeve, first coat the damper nose with a light film of RTV sealer. Position the damper front-side down so its hub is firmly supported. Set the sleeve squarely over the damper nose and, with a wood block against the sleeve, tap the sleeve straight on. Check to make sure it starts straight. After the sleeve is flush with the end of the damper, wipe away excess sealer.

Install Damper

Apply a light coat of anti-seize compound to the nose of the crankshaft and a coat of oil to the damper seal surface. The anti-seize will make damper removal much easier if the need ever arises. The oil provides seal lubrication on initial start-up.

Fit the damper to the crank so it engages the key. If the damper doesn't slide on by hand, use a soft mallet to tap the damper onto the crank by striking the hub only. Do not hit the outer ring. If it takes more than light force to install the damper, use a damper installation tool or a threaded rod with a large washer and nut to pull it into place. Excess force can damage

To ease installation and ease future removal, apply a thin coat of anti-seize compound to damper ID and oil-seal surface.

the rubber bond between the outer ring and damper hub. The damper should go on far enough so the damper bolt will start with at least two threads engaging those in the crank.

After installing gasket(s) to cover with sealer and applying sealer to block, fit cover to block. If cover is not positioned to block with dowels, loosely install bolts and center cover to crankshaft by installing damper. Tighten front-cover bolts.

Draw the damper on the rest of the way with the bolt and torque it to spec. To keep the crankshaft from turning while you torque the damper bolt, insert a large screwdriver, pry bar or hammer handle between a crankshaft counterweight and pan rail.

OIL PUMP & PAN

It's time for some final thoughts before the oil pump and pan go on: "Should I re-torque the mains? Did I final-torque the rods? How about rechecking the rod numbers and torque." Checking for such things now will keep you from sitting straight up in bed in the middle of the night with the same questions the night before your initial startup. Take the opportunity now to check things in the bottom end you might have questions about.

If it's not already there, position the engine on its back. You'll need the oil pump, its driveshaft, the pump pickup, the pan and whatever is needed to seal the pan to the block. Sealing items include gaskets, seals and sealer. You'll find all but maybe sealer in the gasket set.

Oil Pump Driveshaft—All sorts of end configurations have been used on oil pump driveshafts. The most common are the hex drive and tang drive. And the most common way of driving the pump is with the distributor gear through the distributor shaft. Here, the bottom of the distributor is fitted with the drive gear which is driven by a gear on the cam. But a few shafts are fitted with their own gear.

Regardless of your engine's setup, inspect both ends of the driveshaft. This is especially critical with a hex driveshaft. The corners of the drive end should be sharp. If there is any hint the corners are beginning to round off, replace the shaft.

Install with washer and torque damper bolt. To keep crankshaft from turning use a large screwdriver between block and crank counterweight. If oil pan is in place, use two bolts in damper or flywheel flange and bridge them with a screwdriver or pry bar.

A special note about Ford and Chevy engines: Ford uses a small hex-type shaft fitted with a Tinnerman retainer. Replace this shaft. As for the retainer, do a trial oil pump-and shaft fitting. Check that this clip is positioned high enough on the shaft that it cannot pull out of the oil pump

Install oil pump with driveshaft and gasket. Grease will help seal gasket and hold it in place. Don't use sealer on gasket.

HIGH-PERFORMANCE OIL PUMP SHAFTS

High performance oil pump shafts are available, but these large-diameter versions of the original shafts may actually be weaker than the ones they replace. High performance shafts are larger in diameter, but the ends are the same. Consequently, they won't flex as much when loaded. This is sort of like a car's suspension hitting the bump stops when going over a chuck hole. The result is additional shock load is transferred through the oil pump drive system and the engaging surfaces of the shaft at the distributor and pump. Additionally, the stresses in the shaft are localized and higher because of the rapid section change between the smaller ends and the larger diameter—similar to the transition between the threaded section of a bolt and its straight shank. This increases the likelihood of the shaft breaking at the section change rather than absorbing loads as the shaft flexes along its full length. So think twice about installing a "high performance" oil pump shaft.

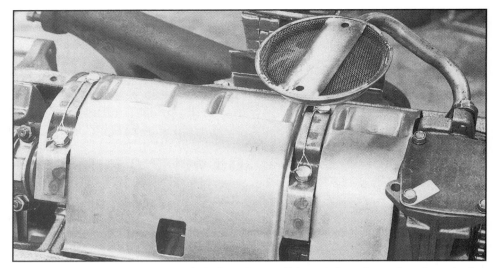

Windage tray installs under crankshaft to keep oil from being churned up by spinning crankshaft, thus reducing horsepower loss. Special main bearing cap bolts are required for installation. Some engines require special main-bearing caps.

when the distributor is removed. As for the Chevy shaft, replace the nylon sleeve that holds the shaft into alignment with the oil pump. If either the Tinnerman retainer or Nylon sleeve fail one way or another, the results can be disaster for your engine. Regardless of your brand loyalty, I'm sure you'll agree that your engine won't last long without oil.

Install Oil Pump

If the pickup is not attached to the oil pump, install it now. It won't be if you got a new pump. It must be transferred from your old pump to the new one. The pickup will bolt to the pump or press into the pump body.

If bolted on, find the pickup-to-pump gasket and use grease to seal it. Don't use sealer, especially RTV silicone. It can squeeze out from the inside, eventually break off and jam the gears. Secure the pickup using the attaching bolts torqued to spec. For a pressed in pickup, position the pump on vise jaws and drive the pickup into place using a section of pipe. The bottom of the pickup should be level with the bottom of the oil pan when installed. Check by loosely installing the pump on the engine. Position the pickup to the pump so it's level and scribe lines on the pickup tube and pump body that align when the pickup will level with the bottom of the oil pan. Before you drive in the pickup, apply some Loctite to the end of the tube. With the pipe butted against the bead near the end of the pickup tube and the scribe marks aligned, drive in the pickup until it bottoms.

Now for installing the pump: You'll need the pump-to-block gasket, the pump, driveshaft and, possibly, pump-to-driveshaft sleeve. Using grease again to seal the gasket, fit the gasket to the bottom of the pump and the driveshaft to the pump. Insert the driveshaft into the distributor-shaft bore and secure the pump to the block. Install the hold-down bolts and torque them to spec. If the shaft has an integral gear, install the pump as described above, but without the shaft. Install the shaft later.

Windage Tray—If your engine uses a windage tray, install it now. If it doesn't, it would be an inexpensive addition that would improve performance, particularly at higher rpm. Check with a dealer to see if windage trays were available for high-performance versions of your engine. You'll need the tray and main-cap bolts that are drilled and tapped for the attaching bolts.

Install Oil Pan

It's time to seal the bottom end of your engine. Find the oil-pan gaskets and/or seals. While you're at it, find the new drain-plug seal and install it. Check the areas around the oil-pan bolt holes. If they are dimpled from the bolts being overtightened, oil leaks will result if the pan is installed as is. Flatten these areas by lightly

Neoprene oil pan end seals in this engine must be worked into grooves in front cover and rear main. RTV sealer is used at joints between front and rear oil-pan end seals and oil-pan-rail gaskets. If yours is a one-piece oil-pan gasket, this is one operation you don't have to be concerned with. Just fit gasket to bottom of engine.

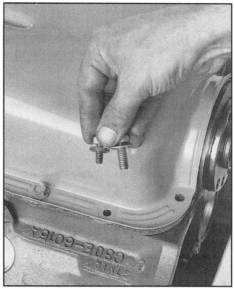

Take a last look at bottom end before you install oil pan. If all is OK, carefully set oil pan on gaskets, being careful not to move them out of place. Install and snug bolts. This pan uses larger bolts at four corners. Don't overtighten bolts and dimple pan flange.

tapping the flange down around the bolt holes. Back up the flange with a punch or short section of round stock clamped vertically in a vise.

Fit the neoprene seals to the bottom of the front cover—sometimes a section of gasket is used here—and to a groove in the rear main-bearing cap. Apply a thin coat of RTV silicone to the bottom of the pan rail, go around each bolt hole and put a dab of sealer where the pan gasket and end seals or gaskets join. Lay the gaskets on the pain rails and fit their ends to the end seals or gaskets as detailed in the instructions in the gasket set. Apply RTV sealer to the front-cover gaskets. Put a little extra sealer where the seals and gaskets join.

Set the oil pan in position, being careful not to disturb the seals and gaskets. Install all oil-pan bolts then snug them gradually. Note that larger bolts may be used at the four corners of the pan to give the additional force needed to pull the pan down over the seals at the front and rear mains. The oil pan bolts will seem to have loosened when you go back to snug them. Because cork gaskets relax and tension in the pan bolts lessen, snug the bolts several times, then torque them to spec. You'll have to do this several times, too, until the gaskets are fully compressed as indicated by the torque wrench.

CYLINDER HEADS

Now that the bottom end is buttoned up, turn your attention to the top end of your engine. You'll need the cylinder heads, gaskets, head bolts and hardened head-bolt washers, if used. Don't forget sealer for the head bolts if they thread into the water jacket. For an aluminum block, you'll need anti-seize compound for the head bolts. A final item is high-temperature aluminum paint or special sealer for the head gaskets. Get it unless the directions in the gasket set say not to use it.

Head Gaskets

Double-check the gasket surfaces by wiping clean the block and cylinder head gasket surfaces with lacquer thinner soaked paper towels. Check for old gasket material, nicks or other imperfections you may have missed and correct any problems.

There should be two solid or hollow dowels for each head. Don't install the heads without them. These are needed to accurately position the cylinder heads on the deck surfaces. If you can't find them, purchase new ones at a dealer, your engine machinist or a junkyard. Being careful not to mushroom or burr them over, tap each dowel in with until it bottoms.

Install Gaskets—Check the head gaskets for damage. Just because they are new doesn't mean they can't be bent or have nicks, cuts or other flaws. Exchange them for another set if you find a problem. A high-quality gasket set is your best insurance against this happening.

Trial-fit the head gaskets to the block decks. Look for marks such as **FRONT** or **TOP** on the head gaskets and install them accordingly. If incorrectly installed, critical water passages will be blocked, resulting in severe cylinder-head overheating. Above all, don't let the looks of head gaskets such as color or sealer bead pattern from one side to the other

Fit head gasket to dowels on deck surface exactly as directed by instructions in gasket set. If FRONT is printed on gasket(s), install that end to front of engine even though printing may face down. Do otherwise and severe overheating may result.

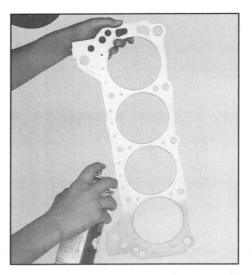

Check instructions in gasket set. If instructions don't say not to, coat both sides of head gasket(s) with an even coat of aluminum paint.

Clean head-bolt threads with a wire brush. Threads in holes and on bolts must both be clean to achieve correct tension from torquing. If more than one length bolt is used, make sure you get them in the right holes.

affect how you install them. It's the alignment of coolant passages that's critical, not what side is up or down.

Unless the gasket instructions say otherwise, apply aluminum paint or sealer such as High-Tack or Copper Coat to both sides of the gaskets. After the paint or sealer dries, lay the head gaskets on the deck surfaces and fit them to the dowels.

Head Bolts—If you haven't already done so, clean the head bolts. The threads must be absolutely clean, so give them the wire brush treatment. And check for length, too. There are almost always two different lengths of cylinder head bolts and frequently three. So clean them, lay them out and count them. You should have all head bolts clean and ready to go before you install the cylinder heads.

Install Heads

Make sure your engine is well supported before you set a cylinder head on it. The pin or clamping bolt should be secured in the neck of the engine stand. If it's not on an engine stand, the engine should be supported with blocks under the engine mounts or pan rails. This is particularly important with a V-type engine. Not only does the cylinder head move the center of gravity up on an engine, the first head on shifts it way off center and out of balance. This can cause an engine to roll over if it's not well supported.

Have at least two head bolts within easy reach. Make a trial run on lifting and setting the cylinder head on the engine. Here's where aluminum heads are an advantage. A finger in an intake port at one end and one in an exhaust port at the other works best for lifting most cylinder heads. Keep in mind that you must hold the cylinder head over the deck surface and move it around so it firmly engages the dowels before you can release it and let go. By this time the arteries in your arms start to feel as though they'll pop. As soon you have the head firmly on the dowels, run two head bolts part way in to keep it in place. If your engine is a V6 or V8,

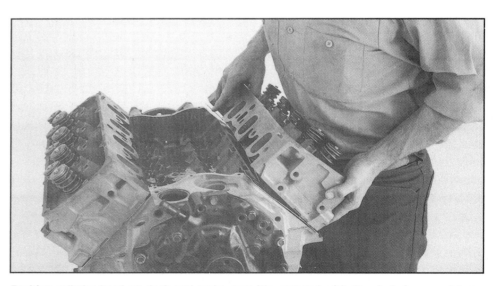

Position cylinder head on deck and make sure it's engaged with dowels before you let go. Fingers in exhaust port at one end and in intake port at other end works well. Thread in at least two bolts to ensure head doesn't fall off.

Before installing, apply RTV sealer to bolts that thread into water passages. If block is aluminum, apply anti-seize compound to bolt threads. If heads are aluminum, install hardened washers under bolt heads. Torque bolts to the specified sequence and value in at least three stages of increasing torque.

set the other head on the block and secure it in the same manner before you install any more head bolts.

If the head bolts thread into an aluminum block, lightly coat their threads with anti-seize compound, oil or apply extreme pressure lubricant to the undersides of the bolt heads and drop them into their holes. For cast-iron heads, lubricate the heads and threads with oil or extreme pressure lubricant. If threaded into water passages, seal the first 2 or 3 threads with RTV. Don't forget the two you threaded in first. Remove them and give them the same treatment. Run the head bolts down, but don't tighten them yet. It's critical that you now follow the head bolt torque sequence for your engine. Do this in 30-ft-lb increments until you reach the final torque value. The typical tightening sequence is from the center outward back and forth, but check to make sure. After final torquing, go back over the bolts in sequence and recheck torques at least once.

At this point you may or may not be finished with torquing head bolts. Typically, they should be re-torqued after initial engine run-in. To determine whether this is necessary, refer to the gasket installation instructions. Head gaskets such as blue Fel Pros are a *no re-torque* gasket. This means you don't have to pull off the valve covers to re-torque the head bolts, which is worth the "price of admission" in my book.

Dig up the old set of spark plugs and install them finger tight. Don't use new ones. Chances are some will get broken, but who cares if they're old? They will keep dirt or any other debris out of the combustion chambers and mask the plug holes for when you paint the engine. Remove the old plugs and install new ones once the engine is in place.

VALVETRAIN

Round up the valvetrain components: valve lifters, pushrods, rocker arms, their fulcrums or pivot shafts and the attaching hardware. You'll also need a squirt can of oil, maybe a can of oil and some moly lube.

Install Valve Lifters

If you are using hydraulic lifters, prime them to prevent a lot of clatter on startup. You can do this one of two ways: force oil through the hole in the side of the lifter until it comes out the center of the pushrod seat; or, place the lifter upright in oil so the hole in the side is submerged. Work the plunger up and down with a pushrod so oil is drawn into the lifter.

Apply moly lube to the lifter foot, smear oil on its side and slide the lifter into a lifter bore. When they are all installed, squirt some oil into the pushrod seat. Remember: If you are using the original lifters and cam, the lifters must be installed on the same lobes. New lifters can go on any lobe.

Install Pushrods

Whether or not you can install the pushrods depends on your engine's intake manifold design. If the pushrods go through clearance holes

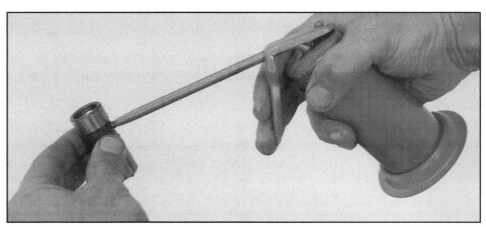

Prime hydraulic lifter by forcing oil through hole in side until it comes out top hole. Lube lifter foot with moly and sides with oil.

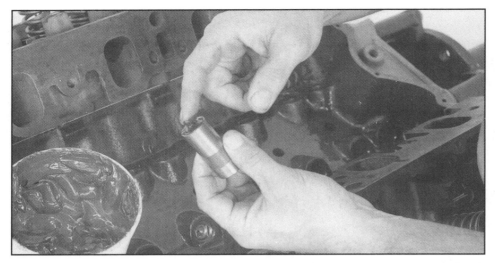

If you're reusing cam and lifters, install lifters in their original bores. If you've lost track of their original order, install new lifters to avoid wiping out a camshaft.

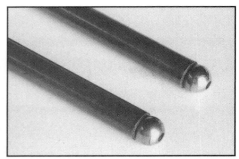

Wear at rocker arm end of pushrod extends farther around ball (top) than at lifter end. If you are installing new rocker arms or didn't keep pushrods or rocker arms in order, swap pushrods end for end.

in the manifold instead of the heads such as that for the big-block Ford or 2.8 Liter Chevy V6, the intake manifold must go on first. In this case install the manifold then continue with installing the valvetrain.

If you haven't already done so, clean and inspect the pushrods. Run a wire through the centers of each and flush it with solvent. Check for debris by holding the pushrod up to a light and looking through it.

New or old, check each pushrod for straightness by rolling it on a smooth flat surface and looking underneath as it rolls. This will quickly reveal a bent pushrod. If a 0.008-in./0.20mm feeler gauge fits underneath the center, replace it.

The rule for installing original pushrods is similar to that for valve lifters. Install old pushrods in order and with the same ends against their rocker arms. Because rocker arm and pushrod contact points wear in together they must be mated when reused. If ball tips are used at each end, determine which is which by comparing their wear patterns. Wear at the rocker arm end extends much farther around the ball tip than does the lifter end due to the higher angular movement of the rocker arm.

When installing new rocker arms, pushrod order doesn't matter. But in this case, reverse the pushrods end for end. Set the ends originally installed in the rocker arms in the lifters. This will have an effect similar to installing new pushrods. As for pushrods that use a ball tip at the bottom end and a socket at the top, they can't be reversed.

Install Rocker Arms

Prepare individually mounted rocker arms by lubricating wear points. Smear extreme pressure grease or moly lube on the pushrod tips and

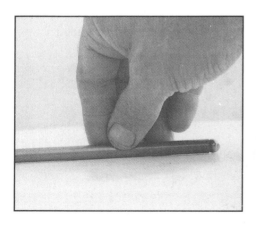

Check pushrods for straightness by rolling on a flat surface. With a light on the opposite side, look under each pushrod as you roll it. If you suspect pushrod is bent, check it with a 0.008-in. (0.20mm) feeler gauge. If the gauge fits, don't try to straighten pushrod. Replace it.

Install and oil pushrods at both ends and at guide plates, if used. Rotate individual rocker arms to side to make room for pushrod installation. Install shaft-mounted rocker arms after all pushrods are in place.

Install shaft-mounted rocker arms on head after all pushrods are in place. Apply high-pressure lube to valve tip and fit pushrods to rocker arms before you begin tightening hold-down bolts. Tighten bolts one turn at a time to keep from shaft.

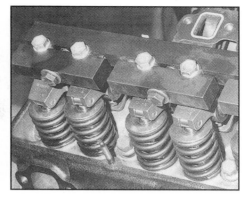
For high performance applications, stud girdle supports individual rocker-arm studs. Adjust valves, then fit girdle loosely to stud extensions. Gradually tighten securing nuts to manufacturer's specs. A "tall" valve cover will be needed to allow room for girdle.

oil the rocker arm pivots. Double-check that the pushrods are still centered in the lifter. Position each rocker arm and pivot on its stud, place the pushrod in the rocker arm pocket, the rocker on the valve tip and thread on the nut a few turns. If used, don't forget to install oil baffles. For rocker arms supported with stands and bolts, set each stand on its boss and run in the bolt a few threads. Finish running down the rocker arm nuts or bolts after all rocker arms are on their pushrods. It's best to do this with the lifter on the cam-lobe base circle. This is impossible, but do the next best thing. For shaft-mounted rocker arms, lubricate the wear points in the same manner as above. Position the crankshaft so none of the lifters on the bank you're working with are at full lift. This will minimize the stress on the rocker arm shaft as you run down the bolts.

Make sure the bolts are in the correct order. Most rocker shaft bolts are different length and one or more may have a reduced shank or other feature that allows oil to pass for top-end lubrication. This is where good organization pays off. Refer to your notes. Getting these bolts in the wrong holes will spell trouble.

Apply extreme pressure lube or oil to the bolt threads and undersides of the heads. Double-check that all pushrods are still centered in their lifters and rocker arms. Set the rocker arm assembly in place and fit the pushrods to the rocker arms. Don't tighten one bolt at a time. Instead, alternate from one bolt to the other until all are snug. Torque them to factory specs.

Adjustable Hydraulic—Once all rocker arms are in place, position the crankshaft on TDC. This will place the cylinder-1 lifters on their cam lobe base circles, or in their lowest positions. Check by rotating the crankshaft back and forth a few degrees to either side. The lifters won't move if cylinder 1 is at the top of its power stroke. If the lifters move it's between the exhaust and intake strokes. If that's the case, turn the crankshaft 360° and realign the TDC mark.

If non-adjustable rocker arms are used, run the nuts or bolts down all the way. If they are adjustable and hydraulic lifters are used, run the nut down until all slack is taken up in the rocker arm and pushrod. Wiggle the

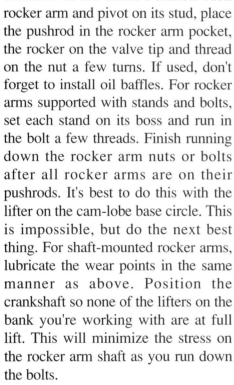

Coat valve tips with high pressure lube. Rotate individually mounted rocker arms so valve end is on tip and pushrod is in rocker arm at other end.

Position crankshaft at TDC firing position to adjust cylinder-1 valves. Lifters will be at their lowest positions in their bores. After you've adjust cylinder-1 valves, adjust all valves using your engine's firing order. Loosely fit distributor and use rotor and cap as reference for positioning crankshaft.

While cylinder is on TDC, rocker-arm nut is tightened to adjust valve. If rocker arm is adjustable with hydraulic lifters, rotate nut 1/2 to 3/4 turn after slack is taken up in pushrod, but before lifter plunger moves. Positive stop nut is simply tightened to spec, but it should not bottom lifter plunger.

rocker arm and pushrod to be sure all slack is gone, but the lifter plunger has not moved down. To position the lifter plunger half way down in the lifter body, rotate the rocker arm adjusting nut about 1/2 or 3/4 turn. Turn the crankshaft 90° if it's a V8, 120° if it's a six or 360° if it's a four and adjust the rocker arms for the cylinder next in the firing order.

For an odd-fire V6, temporarily install the distributor and use the rotor to determine TDC for all cylinders. Find cylinder-1 firing position and mark it on the side of the distributor housing. Mark other cylinder firing positions in the same manner using the distributor cap and engine firing order for reference. Rotate the crankshaft until the rotor aligns with the next cylinder in the firing order, adjust its rocker arms, then rotate the crank to the next position. Continue this procedure until all rocker arms are adjusted.

Adjustable Mechanical—Use the procedure described above to adjust rocker arms used with mechanical lifters. Mechanical lifters run with some lash, or clearance, between the rocker arms and valve tips. To determine the correct lash, check the engine or cam manufacturer's specifications. They'll either specify a cold or hot lash. Regardless, you'll have to do an initial cold lash adjustment so the engine can be started. Do the hot lash once it's up to operating temperature. If you can't find cold lash specifications, use about 0.020 in./0.5mm for the intakes and 0.025 in./0.6mm for the exhausts. These clearances will do for initial startup. Hot lash will have to be set when the engine is at operating temperature.

With the lifter on the cam lobe base circle, insert the appropriate feeler gauge between the valve tip and rocker arm. Wiggle the rocker arm back and forth or rotate the pushrod to check for slack in the valvetrain. Adjust the nut so there's slight drag on the gauge as you slide it in and out.

Non-Adjustable Hydraulic—The only way to "adjust" non-adjustable

On stud-mounted rocker arms used with solid lifters, adjustment can be maintained for longer intervals by installing sheet-metal PAL nuts or Tinnerman nuts on top of standard locknuts.

Check valve tip-to-rocker arm clearance of non-adjustable valvetrain after prying between valve retainer and rocker arm to bleed down lifter.

rocker arms is to use different length pushrods. This will only be necessary if a valve stem extends too far out of its guide due to excessive valve seat grinding or the rocker arm has moved closer to the lifter. In the first case the valve tip should've been ground to compensate for seat work and restore correct rocker arm "geometry." If this was done, no adjustment should be necessary. Excessive cylinder head milling can also cause a problem. In both cases a shorter pushrod is needed. Check the dealership for parts availability. If extensive valve or seat work wasn't done, or the cylinder head(s) or block was not resurfaced, you shouldn't have any problem with valve adjustment.

Two methods can be used to check for correct valve adjustment. Remember that with either method, the lifter plunger should be positioned half way down in its lifter body when the lifter is on the cam lobe base circle, or about 0.045-in./1.14mm travel at the lifter. Using the first method, force the rocker arm against the pushrod to fully collapse, or bleed-down, the lifter. While holding the rocker arm in this position, measure rocker arm-to-valve tip clearance.

There are special tools for rotating a rocker arm against its pushrod. Such a tool hooks onto the rocker arm, allowing you to lever the rocker arm against the pushrod and lifter. This gradually bleeds down the lifter, causing the plunger to fully collapse. Prying with a screwdriver between the spring retainer and rocker arm tip may also work. Better yet, use a pair of long-jaw 10-in. Channellock® pliers. Open the jaws to the widest position, then place the top jaw under the valve end of the rocker arm and the bottom jaw on top of the pushrod end. Lever the rocker arm against the pushrod to compress the lifter plunger.

Be patient. You can't force the lifter plunger down immediately. It will take a minute or more to bleed it down, especially with a new lifter. Once the plunger bottoms in the lifter, hold the rocker arm in that position and check rocker arm-to-valve tip clearance. Clearance should be about 0.100–0.200 in./2.54–5.08mm. Adjust clearance by installing a different length pushrod. If there's not enough clearance, go to a shorter pushrod; if there's too much clearance, go to a longer pushrod.

The next and simplest method of checking lifter-plunger travel is to count the number of turns the nut or bolt takes to go from no slack in the rocker arm or pushrod to when the rocker arm fulcrum contacts its stand, or when the pivot ball contacts the shoulder on the positive stop rocker-arm stud. Once you have all the slack out of the pushrod and rocker arm—the plunger has yet to move—count the number of turns it takes for the nut or bolt to begin to tighten. It should take 3/4 to 1-3/4 turns. Less and the pushrod is too long; more and it's too short; in between and pushrod length is OK.

INTAKE MANIFOLD

With a V-type engine, pour any remaining assembly oil over the valve lifters before you cover the lifter valley with the intake manifold or valley cover. Pour some over the rocker arms, too. Depending on your engine, there will be up to four intake manifold gaskets.

Manifold Gaskets

If a single gasket is used, which will be metal, it also acts as a baffle to shield the bottom of the manifold from hot oil. To ensure sealing is good, it is beaded around the intake ports, coolant passages and both ends at the block. Run a bead of RTV silicone sealer around the coolant passages to make sure they seal.

Sealing—When four gaskets are used, one will go between each head and the manifold and one will seal each end of the manifold to the block. Again, use silicone to seal the coolant passages. To prevent these gaskets from shifting out of position when you install the intake manifold, apply several dabs of adhesive sealer along the length of each gasket. Fit the gasket to the head, lift it up, and refit it until the sealer tacks. As for the end gaskets, don't use them. Not only are they difficult to keep in place when installing the manifold, they typically shrink over time and leak.

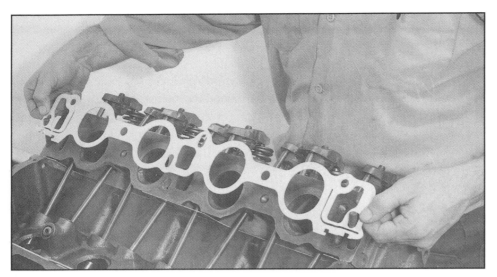
Separate manifold-to-cylinder head gasket is fitted to head. Apply bead of RTV sealer around water passages.

V-type engines use end gaskets between manifold and block. Due to difficulties in sealing, I prefer using a thick bead of RTV at both ends in place of gaskets.

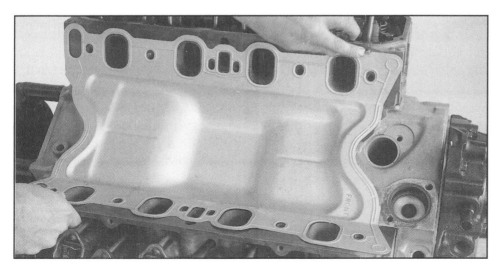
This V-8 uses one-piece metal-shim gasket/baffle under intake manifold.

To eliminate end gasket leakage problems, run a thick bead of RTV silicone across the block at both ends just prior to installing the manifold. But if you insist on using the end gaskets, apply adhesive sealer to the block and seal, lay each seal in position on the block and lift it up as above until the sealer tacks. After you lay the seals in position for the last time apply a dab of RTV silicone at the joints where each seal meets the gaskets at the heads.

Some V-type engines use a separate valley cover to seal the top of the engine. The advantage with this setup is there's an air passage between the intake manifold and engine which keeps the bottom side of the manifold cool. If your engine uses such a manifold, fit the gasket to the cover and install the cover. Next, install the manifold-to-head gaskets with adhesive sealer to keep them in place.

Install Manifold

Set the intake manifold on the engine, being very careful not to disturb the gaskets. To do this correctly, lower the manifold squarely on to the gaskets. It would be a plus if you had a friend to help with this operation. Move the manifold after setting it down and you'll probably shift a gasket or two out of position.

After the manifold is in place, check the gaskets to make sure they haven't moved. If you have any doubts, lift off the manifold and try again. Oil leaking from an end gasket is an aggravation, but a giant vacuum leak at an intake port between the head and manifold can cause engine roughness, oil consumption and spark plug fouling problems.

Install the bolts and, if used, nuts to secure the manifold. Wire brush the threads to make sure they are clean. If you have an aluminum manifold, install hardened washers under the

Be careful when installing intake manifold on a V-type engine. It's too easy to move gaskets out of position. Result will either be an oil leak or a huge vacuum leak and excess oil consumption.

Observe correct intake manifold torquing procedures. Re-torque bolts until gasket(s) are fully compressed.

bolt heads to keep the bolt heads from digging into the soft aluminum. If the bolts thread into a hole that runs into a coolant passage or the lifter valley, use RTV silicone on the threads or a sealing washer under the bolt heads or nuts.

There's no general pattern for torquing manifold bolts, but there is a firm rule: Follow the manufacturer's recommended torque sequence, or pattern, and values for tightening intake manifold bolts/nuts. The typical intake manifold bolt/nut torque sequence follows a zigzag pattern similar to this: Torque the bolts/nuts starting with the middle front right, to the middle rear left, to the middle rear right, the middle front left, out to the ends, back to the middle and so forth. One firm rule is to gradually work up to the final torque in steps of one-third. For example, if final torque is 30 ft-lb, start with 10 ft-lb, go to 20 ft-lb and finish with 30 ft-lb. Re-torque the bolts/nuts in sequence at least two more times to ensure gasket compression is retained.

Tape over the carburetor mounting area with duct tape. This will keep foreign objects such as tools and washers from falling undetected into the manifold. Suffering the consequences of this at startup time is a sure way to ruin your day.

EXTERNAL BITS & PIECES

Valve Covers

Before you install the valve covers, make sure the mounting flanges are clean and straight. If the valve covers are sheet metal—as opposed to cast aluminum—pay particular attention to the bolt holes. If they are dimpled, position the flange over the edge of a bench or vise and hammer them flat. Also check any grommets in the valve cover such as for a PCV valve. Replace any hard, brittle or cracked grommets.

Fit the gaskets to the valve covers. Some manufacturers notch the flanges to hold the gaskets in place for assembly. Tabs spaced around the periphery of a gasket snap into these notches. Another way to ensure the gaskets stay in place is to resort to the old standby—adhesive sealer. The only problem with this stuff is it's tough to get off when and if you need to replace the gasket. But if using adhesive sealer means preventing an oil leak, difficult gasket removal is a minor price to pay. Just don't get carried away. A few spots of adhesive sealer spaced around the gasket flange will do the job.

With the gaskets in place, install each valve covers on the head. Install load spreaders or flat washers under the bolts or nuts and lightly tighten them several times until the gasket is fully compressed. Over-tightened valve cover bolts is a major cause of oil leaks.

Thermostat

The thermostat will either install in the block or intake manifold. More than likely your engine has a wet manifold—coolant flows from the cylinder heads to the manifold on its way to the radiator rather than the block. In this case the thermostat will

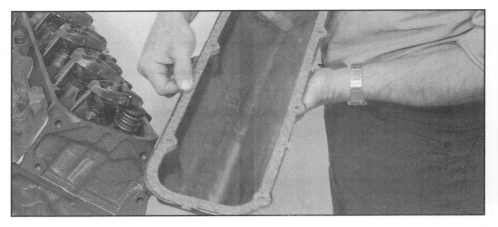

Fit gaskets to valve-cover flanges. Tabs on this cork gaskets snap into notches at edge of valve-cover flange. If holding gaskets in position is a problem, use small dabs of weatherstrip adhesive on flange. When installing valve cover be careful not to overtighten bolts.

Install thermostat in the right direction. Flange fits in recess in block or manifold.

install in the manifold. Otherwise, it will install in the block.

Regardless of where the thermostat installs, the spring end goes toward the engine, or heat source. It should indicate as such on the thermostat. Another caution: Be very careful that the thermostat remains in its counterbore as you install the housing, particularly if the thermostat housing is aluminum. A thermostat that installs horizontally can easily slip undetected out of its counterbore, causing the thermostat flange to be pinched behind the bottom side of the housing. This usually results in the thermostat housing flange to snap off as the bolts are tightened. To prevent this from happening, make use of adhesive sealer again. Run a small bead around the thermostat flange that bottoms in the counterbore. Do the same to the gasket. After the sealer tacks, insert the thermostat in the counterbore. It should stay there. To make doubly sure it stays put, fit the gasket over the thermostat flange and against the housing mounting face and install the bolts.

Paint Engine

This stage of engine assembly is about where you should paint your engine. If it will be part of an authentic restoration, make sure the color is correct. Also be aware that most factory engines were assembled when painted, so all gasket edges, bolt heads, nuts, washers and engine mounts were painted, too. Even aluminum intake manifolds were usually installed, but not masked off. This resulted in a generous amount of overspray around the periphery of the manifold. Although not pretty, it was factory original. Many other components were installed later such as carburetors, exhaust manifolds, fuel pumps, distributors, ignition coils, spark plugs, sending units and accessories. These items were not painted or were painted another color. So leave these off until after you've painted your engine.

If originality doesn't matter, your options are open. But don't get too ornate with the paint scheme. Generally, the best looking engines are one color and dressed up with custom items such as aluminum or chrome valve covers, ignition wires of a bright color, a performance air cleaner and anodized accessories. Some of the slickest looking custom engines are monochromatic—the same color for everything including many items just mentioned.

Spend most of your painting time on preparation. Paint doesn't stick to oil or grease, so clean your engine all over with spray carb and choke cleaner, brake clean or lacquer thinner. Mask off gasket surfaces such as those for the fuel pump, carburetor and exhaust manifolds. Use a quality engine paint. And when applying the paint, start with a dust coat, then apply several thin coats rather than one thick coat. You'll not have as many runs if any and the paint will be thinner, thus less prone to chipping.

Exhaust Manifolds

Whether you install the exhaust manifolds now or after the engine is installed depends solely on whether the engine can be installed with them in place. In the case of the more compact cast iron exhaust manifolds, it'll probably fit. As for headers—tubular exhaust manifolds—they are another matter. Almost without exception, you'll have to install them after the engine is installed. But to make the job much easier, loosely install them to the exhaust system before you install the engine. There are too many variations to say this always works best, but it'll be easier if it does work. Access to the bellhousing bolts and starter will be the biggest problem.

If cast-iron exhaust manifolds are used, check them for cracks. Pay particular attention to the flanged

Check exhaust manifold for cranks. A good welder can make the repair. Smooth mounting faces with a large flat file. Replace header-pipe mounting studs if they are in bad shape.

To make installation of mechanical fuel pump easier, turn crank until fuel pump eccentric is at lowest position. Apply RTV sealer to both sides of gasket.

areas. Some exhaust manifolds are more prone to cracking than others. You'll find out if you have to search for a replacement. Not only will it be more difficult to find, the price will reflect its rarity. Check also the exhaust pipe mounting studs. Replace them if the threads are in bad shape.

Rather than using the gaskets supplied in the gasket set, I recommend that you use RTV silicone for sealing both cast-iron or tubular manifolds to the heads. Sealing is superior with both. An exception would be in the case of a shim-type metal gasket that doubles as a spark plug shield. Use these to provide shielding for the plug wires.

Find the manifold mounting hardware. Typically, different length bolts are used. Some engines are equipped with lifting lugs bolted on with the exhaust manifolds from the factory. If so, at least two bolts per side may have studs projecting from the heads for mounting the plates. If your engine had lifting lugs, be sure to reinstall them. Not only are they useful for lifting your engine, lifting lugs provide originality if you are doing an accurate restoration. The same goes for spark plug shields and metal locking tabs for the manifold bolts. They restore originality and are functional.

Apply RTV silicone sealer to the manifold mounting faces and install the manifold. Tighten the bolts in increments from the center out. Lock the manifold bolts or nuts with locking tabs or stainless steel safety wire. If you use wire, the bolt heads must be cross-drilled for that purpose. At the header pipe, use a new doughnut or gasket. Nothing is more irritating than to fire up a new engine and hear the spit, spit of an exhaust leak.

Water Pump

It takes an enormous amount of work to replace a water pump on an installed fully dressed engine. Because of this I recommend that you replace it now while the job is relatively easy.

If the pump has a sheet metal backing plate, install a gasket between it and the pump housing with silicone sealer. Install the pump to the engine, sealing it with a gasket(s) and sealer.

Fuel Pump

Trial fit the fuel pump. If the pump lever is on the high portion of its eccentric, rotate the crankshaft so the lever is on the low portion of the eccentric and the pump moves in against its mounting surface.

Using RTV silicone sealer, seal the gasket to the pump. Run a bead of sealer around the engine side of the gasket, too. Prelube the pump actuating arm with extreme pressure grease, then install the pump.

Miscellaneous Hardware

An engine can have all sorts of hardware and components fitted to it, especially those from the late '70s on up. Here's where keeping good records is invaluable. I hope you took good notes. There's just not enough space in this book to list the complexities of one brand, let alone all possibilities. Examples, though, include vacuum fittings, ported vacuum switches, divorced choke coils, water temperature sending units, oil pressure sending units, heater hose nipples and EGR valves. If gaskets are needed for sealing, you'll find them in your gasket set. In the case of pipe threads, wrap them with Teflon tape before installing. ■

CAMSHAFT THEORY 7

The question of what camshaft to use always seems to come up in discussions about building engines regardless of whether the engine is for mild street performance or all-out racing. That's why I wrote this chapter; to peel away much of the veneer concerning the "mysteries" of camshaft design so you can make an informed choice, or at least hold your own in these "bench racing" discussions.

Starting at ground zero, the sole purpose of a camshaft and valvetrain is to open the intake valves to admit a combustible air/fuel mixture to the combustion chambers and open the exhaust valves to release the spent air/fuel mixture. This, of course, is an oversimplification. You must also know how much to open the valves and at what rate, when to open them and when to close them.

Knowing how a camshaft relates to valve action and the resulting effect on performance, driveability, mileage and emissions will allow you to make a better decision when choosing a camshaft. Fortunately, manufacturers have made this job relatively simple. Most list in their catalogs cams according to application, such as those for towing, street/strip, strip only and so on all the way up to all-out racing. They reference important factors for specific engines such as vehicle weight, gearing, rpm range and transmission type. Your responsibility is to realistically and accurately supply such information in order to make the best choice.

Big-block drag engine has 0.650-in. valve lift, way too much for even the most radical street-performance engine. Always err on conservative side when choosing a cam regardless of application. Although camshaft manufacturers list specific cams for specific applications in their catalogs, you must still give them the correct information. An understanding of the basic relationship between camshafts and valve timing is essential.

Other factors to consider when choosing a cam is whether to use the original valvetrain setup or switch from hydraulic lifters to solid lifters or from flat to roller followers. And always keep in mind that the most common error is to "overcam" an engine, or install a camshaft that is "too big"—too much lift and duration. The consequence is similar to that of over-carbureting an engine—you'll end up with a "lazy" engine, or one that does not perform well for its intended use. If you still have difficulty with choosing a cam, manufacturers have customer-service lines manned by experts that can help.

Even though manufacturers catalog camshafts according to application, you should understand technical terms that are commonly described by simple terms such as a *big* camshaft or a *bigger* camshaft. Such

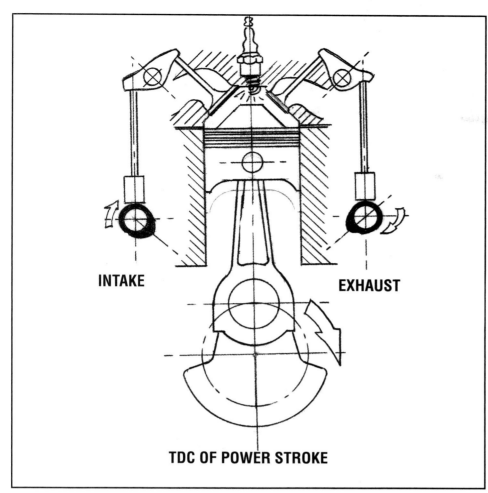

TDC of power stroke: Intake and exhaust lobes are separated for clarification. Both valves are closed and the air/fuel mixture was ignited several crankshaft degrees before piston reached TDC.

descriptions make it sound as though camshafts are chosen like you would choose a box of popcorn. Ask someone who uses such terms to explain exactly what they mean, then be ready to hear a lot of nonsense or nothing at all.

To fully understand terms such as *lift*, *duration* and *overlap* and the effect of each on engine performance, you must first know what goes on in a cylinder during each event of a four-stroke-cycle piston engine.

THE FOUR-STROKE CYCLE

Four separate and different strokes of a four-stroke engine must occur for each piston to complete one cycle so power can be produced at the crankshaft. As for what one stroke means, it's the degrees a crankshaft turns as the piston moves between the top of its bore—top dead center (TDC)—and the bottom of its bore (BDC) in either direction. The crankshaft rotates 180° for each of these strokes. To complete one full cycle—four strokes—two complete crankshaft rotations must occur.

The four strokes of the piston engine are named and duplicated . . . and always occur in the same order for each complete cycle. In any order the sequence of strokes is always: INTAKE, COMPRESSION, POWER and EXHAUST. The cycle is complete and ready to start the next cycle in the same order of strokes. We can start with any stroke, but for clarity let's start at the top of the power stroke:

Many would have you believe that at TDC of the power stroke, the air/fuel mixture explodes with a BANG! Then at BDC of the power stroke the exhaust valve instantly pops completely open and remains there until the crank rotates 180° and slams shut when the piston reaches TDC. The plot thickens: They would also have you believe that at the end of the exhaust stroke the intake valve pops open fully in an instant just as the exhaust valve slams closed. To complete this fairy tale, they would also have you believe that the intake valve closes instantly when the piston reaches BDC of the intake stroke. Don't believe it.

Although there are actions and reactions that are fast, nothing happens instantly—nothing. Even the *burning*—not exploding—of the air/fuel mixture takes time, or about 0.003 seconds from the time ignition occurs. The same goes for the movement of gases through an engine. They don't react instantly anymore than a car can, just quicker. Such things as acceleration, deceleration and inertia are necessary parts of the operational picture of the four-stroke engine.

Overlap—Valve and piston movements *overlap* one another—one occurs before or after the other rather than at the same time. So as the piston reaches TDC or BDC, a valve will have already opened or closed. Taking this further, valve movements also overlap—both intake and exhaust valves will be open at the same time. This is the overlap referred to in camshaft specifications. The amount piston-to-valve or valve-to-valve overlap must vary according to engine design and requirements. And what works best in one engine typically

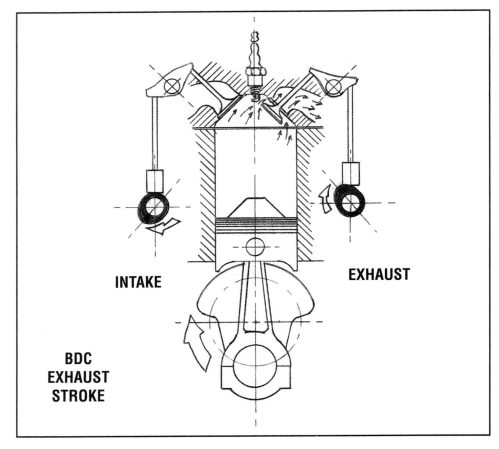

BDC of exhaust stroke: Exhaust valve opened before piston reached bottom center of power stroke to blow down cylinder pressure. Exhaust valve will not close until after TDC of intake stroke.

won't be the best in another engine. Let's dissect each of the four strokes to bring all of this into a clearer perspective.

Power Stroke

Starting at exact TDC of the power stroke, both valves are closed as the piston pauses at the top of the bore in preparation of being forced back down by increasing cumbustion chamber pressure. Ignition of the air/fuel mixture was initiated a few crankshaft degrees before TDC, starting a fire that propagated out from the spark plug above the piston as it passed TDC, thus creating a rapid but uniform pressure and temperature rise.

Maximum pressure should not be reached at TDC because it would tend to push the piston, connecting rod and crank out of the bottom of the crankcase. Instead, maximum pressure occurs about 15° after TDC. The piston continues to be forced down the bore toward BDC by pressure from the burning air/fuel mixture. The leverage the piston and connecting-rod assembly exerts on the crankpin forces the crankshaft to rotate, resulting in what we express as *torque*.

Before the piston reaches BDC of the power stroke, the exhaust valve starts to open. Most of the usable pressure exerted by the still burning and expanding air/fuel mixture was transmitted to the crankshaft through the rod-and-piston assembly. To take advantage of this situation, the exhaust valve starts to open, allowing the remaining pressure in the cylinder to *blow down*. Most of the pressurized gases can now begin to escape past the exhaust valve and into the exhaust port. By the time the piston reaches BDC, the exhaust valve is farther off its seat, albeit not yet fully open. This blow-down process reduces much of what would otherwise be pumping losses. Rather than using the piston to push out combustion gases during the exhaust stroke and wasting power, residual pressure in the cylinder blows them out.

The power stroke, punctuated by the blow-down period, ends when the piston reaches BDC. The crankshaft has now rotated 180° from TDC.

Exhaust Stroke

The official exhaust stroke begins at exact BDC of the power stroke even though exhaust gases started leaving before reaching BDC. This is the first of four periods of each cycle where valve motion overlaps piston position.

The instant the piston reaches BDC it starts to push the remaining exhaust gases past the opening exhaust valve. The remaining pressure in the cylinder pushing against the piston now resists crankshaft rotation, thus creating the dreaded power-robbing *pumping loss*. The result is less power at the crankshaft, but not as much had it not been for the blow-down period.

Well before the piston reaches TDC on the exhaust stroke, the exhaust valve has reached maximum lift and has begun to close. A lot of things begin to happen. Upon approaching TDC of the exhaust stroke, the intake valve starts to open. This is the second of four periods during a cycle that valve motion overlaps piston position. It is also the beginning of *valve overlap*—both exhaust and intake valves are open. The exhaust valve is closing and the intake valve is opening.

The piston slows as it nears TDC of the exhaust stroke. It naturally follows that the departure of exhaust gases from the cylinder slows too, making it virtually impossible for all to leave past the exhaust valve before it closes. And because the intake valve is opening, some of the exhaust gases make an about face and head for it as their only escape route due to lower pressure at the intake valve. Unfortunately, exhaust gases backflowing into the intake port dilute the incoming air/fuel mixture. This is unavoidable because exhaust gases have some weight (technically mass), thus inertia, so the gases flowing into the intake port resist changing speed or direction.

The piston completes its second stroke. During this stroke, the crank has rotated 1/2 revolution, or another 180° for a total of 360°. The four-stroke cycle is half-finished. Even though piston speed in the immediate vicinity of TDC is relatively low there's vigorous activity in the combustion chamber, around the valves and in the ports.

Intake Stroke

The instant the piston stops at the top of the exhaust stroke, the intake stroke begins. The piston starts its trip back down the bore. Crammed behind the intake valve in the port is the air/fuel mixture waiting for the valve to open far enough to let it begin flowing. The exhaust gases that entered the intake port on the exhaust stroke don't get too far before they are overcome by the onrushing air/fuel charge on its way to the cylinder.

Overlap Breathing—Backing up a little, intake valve opening begins before TDC of the exhaust stroke, reducing pumping losses similar to the blow-down action on the power stroke. In this case, though, cylinder pressure and gas-flow directions are reversed. Because both intake and exhaust valves are partially opened at TDC between the exhaust and intake strokes, a phenomena known as *overlap breathing* occurs. How much of this occurs depends on the engine, speed at which it's turning, throttle opening and the intake and exhaust systems. It's not so much an "exchanging of breaths" between exhaust gases and the air/fuel mixture, but how much fuel can be forced out the exhaust port before the exhaust valve closes.

Overlap breathing occurs in all engines to some degree. A lot of fuel forced out the exhaust in an engine built for racing seems to indicate the volume above the piston has been forcibly scavenged, or cleared of all exhaust gases. Not exactly true. What more likely happens is the heavier component—fuel molecules—in the air/fuel mixture resist changing direction due to higher inertia. Consequently, some fuel ends up going out the exhaust. On the other hand, the lighter component—air molecules—along with some exhaust gases change direction more easily, so less air exits the exhaust system. Overlap breathing then seems not to be an accurate description of what happens. Rather, it implies there's a single path of flow from the intake port to the exhaust port. In the real world, a high percentage of gas particles separate from the air/fuel mixture and goes out the exhaust as a rich air/fuel mixture and some exhaust particles reverse direction and get trapped in the combustion chamber.

To complicate things further, some of the air/fuel mixture in the combustion chamber escapes burning in spite of the advancing flame front

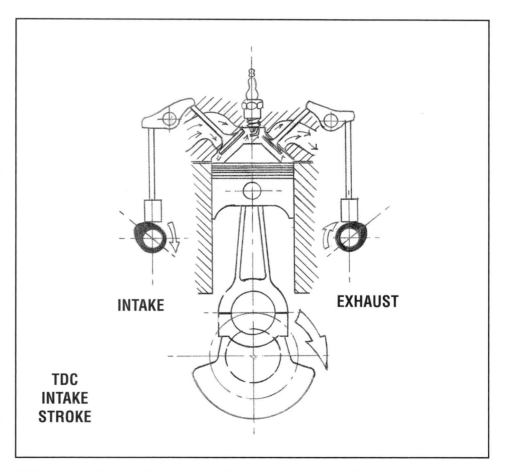

TDC of intake stroke: Both valves are off their seats, exhaust valve closes as intake valve opens during this valve-overlap period. As overlap is increased, dilution with exhaust gases of air/fuel mixture occurs at lower engine speeds as does loss of some air/fuel mixture past exhaust valve.

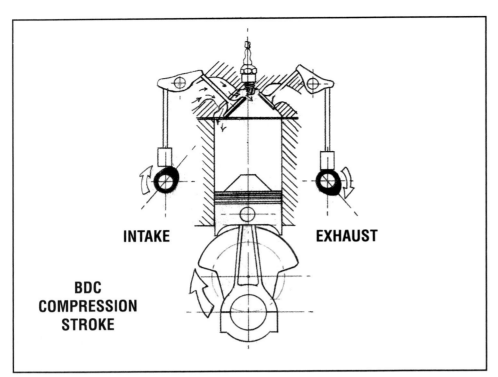

BDC of compression stroke: Intake valve is held open after BDC to profit from inertia of the air/fuel mixture and resulting filling of the cylinder.

and high cylinder temperature, pressure and turbulence. Unburned particles are found in remote areas of the combustion chamber such as between the top of the piston and cylinder wall down to the top ring and in other remote areas. This is because the turbulence of the flame front has calmed and cooled in these locations, preventing the ignition of these particles. Furthermore, because these particles are heavier and less active than exhaust particles, they are among the last to leave the combustion chamber. After the piston passes TDC of the exhaust stroke to begin the intake stroke, some unburned particles do find their way past the closing exhaust valve, and some don't.

The next significant event to occur after TDC of the exhaust stroke is the beginning of the intake stroke. The closing of the exhaust valve follows shortly afterwards. This is the third time valve motion overlaps piston position. As the piston begins its journey down the bore to BDC, the intake valve continues to open. Long before it reaches BDC the intake valve begins to close.

Pressure—As the piston travels down the bore on the intake stroke, vacuum—a pressure differential—is created in the cylinder. How much vacuum there is partly depends on how much the intake valve is open as the piston starts down. The more the valve opening, the less the vacuum—pressure in the cylinder is closer to atmospheric—and vice versa. Here, again, is a pumping loss. The lower pressure above the piston resists its downward motion. This is a necessary evil. If a pressure differential didn't exist between that in the cylinder and that in the port, the air/fuel mixture would not fill the pressure void above the piston. As indicated earlier, the air/fuel mixture starts to enter the combustion chamber before the piston reaches TDC on the exhaust stroke if conditions are favorable.

With a naturally aspirated engine—no turbo- or supercharging—the air/fuel mixture is forced into the cylinder by atmospheric pressure. The air/fuel mixture is not drawn into the cylinder, it isn't sucked in nor does it fall in. It is pushed in by atmospheric pressure in an attempt to fill the volume of low pressure created in the cylinder by the piston moving down the bore. Cylinder filling continues as the piston moves to BDC . . . and beyond.

At BDC of the induction stroke, the third stroke of the four-stroke cycle is completed. Another 180° rotation of the crankshaft has occurred for a total of 540°.

Compression Stroke

Again, the piston reverses direction as it heads back up the bore to TDC on the compression stroke. The intake valve is closing but it's not fully closed. Similar to its movement immediately before and after TDC, piston movement in the vicinity of BDC is "lazy." The crankpin moves through a wide arc before the piston moves significantly toward TDC. Regardless, cylinder filling continues because inertia of the air/fuel mixture has more effect than the upward movement of the piston—at least for a while.

Inertia—About this "inertia thing". compare it to a car attempting to stop or turn at high speed. The higher the speed or weight of the car, the more difficult it is to do. Just look at inertia as the resistance an object has to changing speed and direction. Inertia is what causes anything with weight, including air, fuel and exhaust gases, to maintain speed and direction. That is why the air/fuel mixture resists speed and direction changes as it fills the cylinder as the piston begins to rise after BDC. This goes on until the piston forces the inflowing air/fuel mixture to slow, stop or even reverse direction. This is where closing the intake valve at the precise time becomes critical. The challenge is to close the intake valve before the air/fuel mixture stops flowing into the cylinder.

Occurring long before the piston reaches TDC on the compression stroke, intake valve closing is the fourth and last time valve motion overlaps piston position. From now until the exhaust valve opens near the bottom of the power stroke, both intake and exhaust valves remain closed.

The air/fuel mixture that's trapped in the cylinder is the energy that will be released during the power stroke to turn the crankshaft through the remaining strokes and to supply usable power at the crank flange. This process begins during the compression stroke as the rising piston compresses the trapped air/fuel mixture. The air/fuel mixture is also heated and vigorously agitated by the upward movement of the piston. Because heated air particles try to expand, cylinder pressure is increased further. All this occurs as cylinder volume decreases as the piston approaches TDC. During this frenzied motion, the fuel particles are separated and vaporized by the same compression heating, bringing them into closer contact with the air particles. This violent turbulent mass is now ready for . . . ignition!

Ignition

It is time to light the fire. At about 30° before TDC of the compression stroke, several thousand volts of electrical energy are delivered to the spark plug, causing a spark to jump between the center and ground electrodes. The heat energy of the spark ignites the air/fuel mixture closest to the spark plug electrodes. A chain reaction is set off that will consume most of the air/fuel mixture in the combustion chamber. As the flame front advances from the point of ignition, it moves in a more-or-less uniform manner to force the unburned mixture into an even higher degree of turbulence, causing rapid cylinder temperature and pressure increases.

This early stage of ignition goes on as the piston passes TDC of the compression stroke. At this point the piston completes the last of the strokes. Another 1/2 revolution of the crankshaft is complete, adding 180° of crankshaft rotation to the previous 540°. Total crankshaft rotation is now 720°, or two complete crankshaft revolutions. Theoretically, the compression stroke is complete as is the entire cycle. But what really happens?

It should now be obvious that the strokes overlap each other. The power stroke began when the air/fuel mixture was ignited *before* the compression stroke was over at TDC, the exhaust valve opened *before* the power stroke reached BDC, both valves were opened on both sides of TDC of the exhaust stroke . . . the finishing phase of one stroke. And the finishing phase of one cycle is the beginning of another cycle no matter which stroke it starts with.

CHANGING VALVE EVENTS

To dig further into the four-stroke mystery, let's look at how changing cam lobe *profile*—opening and closing valves at different points—changes the performance of an engine.

Exhaust Valve Opening

Early Exhaust Opening—What happens if the exhaust valve is opened earlier than normal? Starting the blow-down period earlier can be beneficial for an engine that's operated at higher-than-normal engine speeds. This is because most of the useful cylinder pressure is used up somewhere between 80° and 90° before BDC of the power stroke. On the negative side, very early exhaust-valve opening can reduce power output, particularly at lower engine speeds.

Early exhaust valve openings are usually the beginning component of longer durations—valve open periods—in race engines. The bottom line is, exhaust valve opening points that are kept within reason have the least effect on engine performance than exhaust valve closing points or opening or closing points of the intake valve.

Late Exhaust Opening—Later exhaust-valve opening—piston is closer to BDC of the power stroke—helps low rpm performance by keeping pressure on the piston longer. And at higher rpm, power may or may not be affected by this reduced blow-down time. Very late exhaust valve opening points correctly combined with other valve events can reduce exhaust emissions by retaining cylinder heat, allowing it to dissipate before releasing it from the cylinder. This reduces oxides of nitrogen (NOx), carbon monoxide (CO) and unburned hydrocarbons (HC)—gasoline. It can also improve engine performance through a broader rpm range if correctly done.

Intake Valve Opening

The next sequence of events—intake valve opening—generally excite performance enthusiasts most. Early intake valve opening is half of the overlap story. Opening the intake valve early—piston farther away than normal from reaching TDC on the exhaust stroke—causes the engine to be rough and balky at low engine speeds. This is because of greater

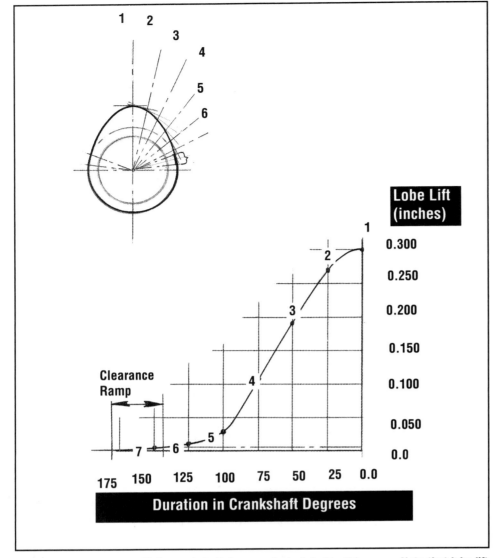

Camshaft profile, or lobe lift in crankshaft degrees, is depicted by this curve. Note that lobe lift starts at point 1, which is the centerline of the lobe, and at zero crankshaft degrees.

higher rpm may be a reasonable trade-off for better low- and mid-range performance, better idle and off-idle characteristics, and possibly improved fuel economy provided the right foot—the one over the throttle—can be controlled. Further, very late intake valve opening points can result in reduced exhaust emissions without an accompanying loss in performance.

Exhaust-Valve Closing

The other end of the valve overlap period is marked by the exhaust valve closing point. Late exhaust valve closing—piston farther down on the intake stroke—contributes immensely to evil low-rpm performance. This is because the air/fuel mixture can keep on going through the combustion chamber and escape out the exhaust where it does nothing useful except help cool the closing exhaust valve. Instead, it should get trapped in the combustion chamber where it can do useful work and not increase HC emissions.

High valve overlap allows race engines to tolerate high compression and full spark advance at low engine speeds. Peak cylinder pressure is relatively low due to a substantial loss of the air/fuel mixture out the exhaust and a dilution with exhaust gases of the trapped air/fuel charge. This prevents abnormal combustion—detonation—from occurring. As rpm increases, late exhaust valve closing allows a higher percentage of the exhaust gases to exit the cylinder because increased inertia keeps them going in that direction. This results in increased cylinder pressures and power output. But there remains some "bleed-off" of the air/fuel charge, which could limit maximum power output.

Late Exhaust Closing—Late exhaust valve closing, either by itself or combined with early intake valve

exhaust gas dilution of the air/fuel mixture as it attempts to enter the cylinder. Remember overlap breathing?

As engine speed increases, flow efficiency to and from the cylinder improves. Velocity and the resulting inertia of the mixture at higher rpm overcomes most exhaust gas dilution to improve power output. But excessively early intake valve opening points hurt power at low- and mid-range engine speeds, making power and response acceptable only at high rpm. Early intake valve opening points are associated with long intake valve duration—valve-open time—used in race engines.

Late Intake Opening—Conversely, later intake-valve opening—piston is closer to TDC on the exhaust stroke—gives smooth engine operation during idle, off-idle, and at low- and mid-range engine speeds. Additionally, manifold vacuum is retained for operation of vacuum accessories such a brake boosters providing the other three valve opening and closing points are reasonable. Although some power may be sacrificed at higher engine speeds, a slight loss of power at opening, hurts low- and mid-range performance, driveability, fuel economy and emissions. And very late exhaust-valve closing could result in a reduction of maximum power.

At the beginning end of the show, earlier exhaust valve opening results in a smoother operating engine, but not necessarily at the expense of performance. Engine operation may not only become more civilized at lower rpm, top-end power may improve. Overall engine flexibility is enhanced, particularly if earlier exhaust valve opening is combined with later intake valve opening. Opening the exhaust valve very early can also reduce exhaust emissions and, if other engine systems work together, improve performance.

Intake-Valve Closing

In order of occurrence, the exhaust valve was opened, the intake valve was opened and the exhaust valve was closed. Although intake-valve closing is the last valve event to occur, it is the one that can make or break the performance of an engine. Where the intake valve closes in relation to piston position has more effect on engine operating characteristics than the previous three opening and closing points combined.

Late Intake Closing—When optimized for a race-prepared engine, late intake valve closing can capture within the cylinder a larger volume of the air/fuel charge before pressure reversal caused by the rising piston forces it back past the closing intake valve. Just where the most desirable intake valve closing point is in an engine is difficult to determine. Although later intake valve closing may be acceptable, even necessary at high rpm because of the reason just stated, it is self-defeating due to increased reverse pumping as engine speed drops. Conversely, the reverse pumping action is either killed outright or significantly reduced by increasing engine speed and the effect of the direction and inertia of the tail end of the air/fuel charge entering the cylinder.

Extremely late valve closing makes the effect of reverse pumping worse. It can get so bad that mid-range response will be so poor that the engine won't be able to pull itself into its favorable power-producing speed range. Late intake closing also increases exhaust emissions. A balancing act goes on in an effort to avoid coming up with a "peaky" engine—one that operates best at a high but narrow rpm range. An engine that has the intake closing point delayed too long will have a driver that is accustomed to cars going by when coming out of corners or off the line. By the time the engine is making "good power" it's time to get on the brakes.

Earlier intake valve closing is what gives an engine power at the broadest speed range. It also improves idle, off-idle and part-throttle performance . . . about everything desirable. This is because minimal reverse pumping takes place at lower engine speeds, allowing more of the air/fuel charge to be trapped in the cylinder and put to work. Earlier intake-valve closing even reduces exhaust emissions, improves fuel economy and may also increase performance.

There is one factor that you must deal with when it comes to valve timing . . . regardless of application: *Compromise.* It's always a compromise between what you want versus what you can get. It's always a trade-off. But "overcamming" an engine is the easiest trap to fall into, resulting in a lazy, soggy performing engine that gulps gas and blows out emissions. Avoid it.

Piston & Crankpin Position

Now that we've looked at the basic relationship of valve events and piston position relative to TDC and BDC, let's dig a little deeper. To get a clearer picture of what's going on, let's look at the positions of the crankshaft and piston between these two points.

Piston Velocity—The angular rotation of the crankshaft within all practicality is constant at any given engine speed. But due to the relationship between the crankpin, piston and the connecting rod, piston motion is not constant. Rather than having constant motion, the piston accelerates and decelerates in between stopping and starting twice to change directions for each crank revolution, once at TDC and once at BDC.

Now for a surprise: The crankpin is not 90° from TDC or BDC when the piston is midway between its up or down stroke. Instead, it's less than 90° down from TDC at mid stroke. When the crank throw is at 90° the piston travels more than half stroke from TDC. The plot thickens: Because the piston travels farther during the "first" 90° of crankshaft rotation, piston velocity is higher from TDC to the midpoint than it is during the last half of crank rotation. This is simply because the piston has more distance to travel for the same amount of crank rotation for the first 90° of crankshaft rotation. Further, maximum piston velocity occurs before 90° of crank rotation from TDC. This happens exactly when the centerline of connecting rod is 90° to a line drawn between the centers of the crankpin and main bearing journals. Maximum piston velocity also occurs at the midpoint of piston travel. The crankshaft angle at which both occur is found by calculating the tangent of the angle, which is equal to connecting rod length divided by

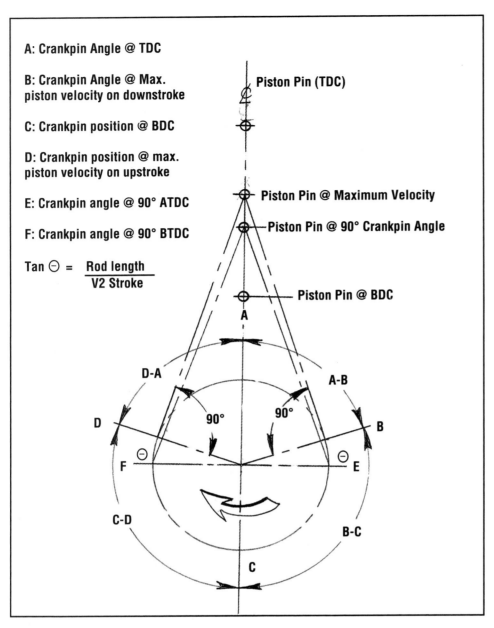

A: Crankpin Angle @ TDC

B: Crankpin Angle @ Max. piston velocity on downstroke

C: Crankpin position @ BDC

D: Crankpin position @ max. piston velocity on upstroke

E: Crankpin angle @ 90° ATDC

F: Crankpin angle @ 90° BTDC

$$\tan \Theta = \frac{\text{Rod length}}{1/2 \text{ Stroke}}$$

Where the crankpin is at TDC and BDC is fairly obvious, but not so obvious when the piston is halfway down the bore. The crankpin doesn't rotate 90° from TDC, it rotates less than 90°. How much a crank rotates from TDC to when the piston is halfway down its bore depends on crankshaft stroke and connecting-rod length. The angle is calculated using the above equation.

crankpin radius, or 1/2 crankshaft stroke.

The drawing above illustrates a typical engine. It has a 4-in. stroke and a 6-in.-long connecting rod. From the above equation, maximum piston velocity is reached at a crankshaft angle whose tangent is 6.000 in. divided by half the stroke, or 2.000 in. The result is Tangent Θ = 3.000. Using your trusty calculator, Θ = 71° 34'.

The piston accelerates from TDC until it reaches this angle. From this point of maximum acceleration until it decelerates to a stop at BDC, the crankshaft must travel another 108° 26', or 180° — 71°34'. This is true for both power and intake strokes. On the other side it takes 108° 26' for the piston to accelerate from BDC to its point of maximum velocity on the exhaust and compression strokes. The piston then decelerates on its trip up the bore until it stops at TDC for 71° 34' of crankshaft rotation.

Alternate, but unequal periods of acceleration and deceleration exert considerable influences on optimum valve timing for given applications. Different rod lengths or strokes change how a piston moves up and down its bore. And because the piston is doing the pumping, valve opening and closing must be tailored to accommodate the piston.

Valve Timing on Piston Upstroke

Let's look at how the valves and piston should "dance" with each other starting with the three valve events that occur during the upstroke of the piston.

Exhaust Lift—Maximum lift should be reached approximately when the piston reaches its maximum rate of acceleration on exhaust stroke to minimize pumping losses.

Intake Open—The intake should begin to open about where the piston reaches its maximum rate of deceleration during the exhaust stroke. This is to prevent excessive air/fuel-mixture dilution as the front of it enters the combustion chamber.

Intake Closed—The intake should close before piston reaches maximum rate of acceleration on the compression stroke to minimize effects of pressure reversal in the cylinder.

Keep in mind that the piston behaves the same in terms of acceleration, deceleration and speed on both exhaust and power strokes. But what the valves are doing relative to the piston is different, and the difference in terms of importance is considerable. The most important by far is intake valve closing. This is followed in importance by intake-valve opening. A distant third is

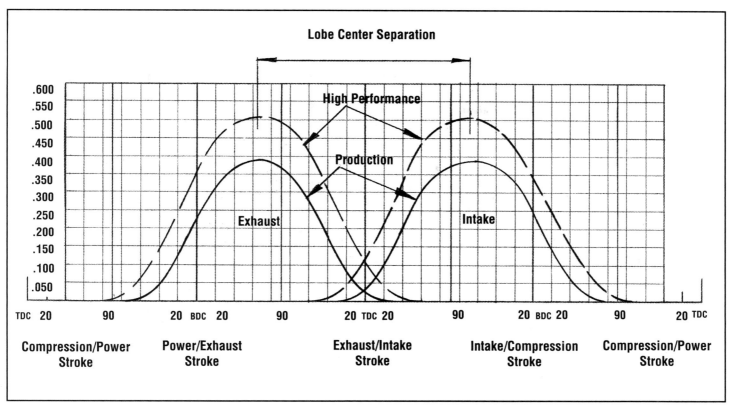

Valve lift curve of typical production cam and compared to aftermarket performance cam. Generally, more area under curve means more power output. Increasing area under curve is done with higher valve lift and duration. Opening and closing valves faster also increases area under curve.

maximum exhaust valve lift.

Valve Timing on Piston Downstroke

As the piston travels from TDC to BDC, it accelerates and decelerates the same on both intake and power strokes. Just as on the upstroke there are three valve events on the downstroke:

Exhaust Opens—Blow-down occurs during piston deceleration at the end of the power stroke. Most of the combustion energy has been converted to useful work, so it's time to get rid of it so that pumping losses are minimized. Cylinder pressure is much higher than atmospheric at this point, so exhaust gases blow out past the exhaust valve. Cylinder pressure is reduced to slightly more than atmospheric when the piston begins the exhaust stroke, meaning the piston has less pressure to push against as it passes BDC and begins its upstroke.

Exhaust Closes—It closes at the beginning of the intake stroke as the piston accelerates, but before it reaches its maximum rate of acceleration. This minimizes the escape of excess air/fuel charge past the exhaust valve and keeps the trailing end of the exhaust gases from being drawn back into the cylinder as pressure changes from above atmospheric to below atmospheric.

Intake Lift—Full valve lift should be reached during piston deceleration, but before maximum rate of deceleration. The intake valve lags behind the piston so a pressure differential is created between the cylinder and atmosphere for a maximum number of crankshaft degrees as is practical. This allows the incoming air/fuel charge to reach the highest possible velocity to maximize its inertia, thus continue cylinder filling as long as possible into the compression stroke.

Of the six valve events just described, some are more important than others. In order of importance, they are:

1. Intake valve closing.
2. Intake valve opening (could be reversed with exhaust valve depending on conditions).
3. Exhaust valve closing (could be reversed with intake-valve depending on conditions).
4. Exhaust valve opening.
5. Maximum intake valve lift.
6. Maximum exhaust valve lift.

SELECTING A CAMSHAFT

As stated earlier, you have to be honest with yourself when choosing a cam. This requires providing answers to some straightforward questions. Easy questions to answer are: What is my engine's displacement, vehicle weight, final gear ratio, transmission

If it's off-roading you're in to, concentrate on maximizing bottom-end torque, not high-rpm horsepower.

type and intended application? Questions to answer concerning performance are tougher because they are more or less subjective.

Determine Your Needs

Before you attempt to come up with the answers, clear your mind of all fairy tales, alignment of the stars, myths and prejudices. The object is to get a camshaft that works for your engine regardless of duration, overlap, lift and such. As an example, a small-displacement engine needs all the torque it can produce, particularly at low engine speeds. Long-duration cams with very high lifts are out of the question. Although great for most race engines, a street vehicle so equipped would need a push from the neighborhood kids to get out of the driveway. With that said, answer the following:

1. What do I want from my engine that it didn't have?
2. Am I willing to sacrifice fuel economy for more power?
3. Is a smooth idle and low-speed driveability important?
4. Are exhaust emissions or laws relating to them a problem?
5. Do I want power at low and mid-range engine speeds or is higher-rpm output more important?

Additional questions address practical matters such as:
6. Will I do the installation or will I farm out the job to someone I know is competent?
7. If needed, am I willing to notch the pistons in order to obtain the correct piston-to-valve clearances?
8. Can I get all the right parts the first time from one source?

Using the answers to the above questions written down on a sheet of paper and a catalog from your favorite camshaft manufacturer, begin the selection process. At the top of your list write DO NOT OVERCAM. "Too much" camshaft is much worse than "too little."

Street Cams

For the street, you're dealing with what is basically a stock engine where idle, off-idle, throttle response, all-around driveability and emissions regulations are important considerations. Realize that except for the "dog-days" of the early '70s through the early '80s where engine performance was nonexistent primarily because of cam design, it's virtually impossible to improve performance at one engine-speed range without compromising other characteristics, including performance at another speed range. But if you want to install something other than the production cam, let's take a look at the numbers: Generally, lift should not exceed 0.470-in. nor more than 240°–270° *effective duration*, which is really *advertised duration*. Effective duration begins when the valve is far enough off its seat to allow the port to flow, or about 0.015 in. This translates to 0.010-in. movement at the lifter with a rocker-arm ratio of 1.5:1.

When comparing camshaft specifications, make sure they are obtained using the same checking method. For example, one cam manufacturer may advertise a cam

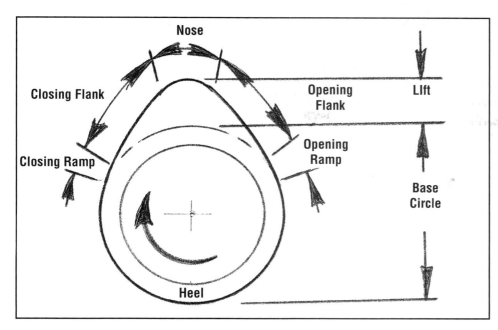

Opening ramp of solid lifter cam lobe provides a smooth transition between an unloaded valvetrain with lash and closed valve to one that is loaded to accelerate the valve to full opened. Closing ramp eases closing valve on to its seat before unloading valvetrain.

that has 300° of effective duration measured at who knows at exactly what lifter movement. The same cam may have a duration of 220° when measured at 0.020-in. lifter movement or 205° duration and no overlap when measured at 0.050-in. lifter movement. The SAE (Society of Automotive Engineers) specifies 0.004 in. of lifter lift for measuring duration. This will result in 40°–50° more duration than if measured at a lifter lift of 0.050 in. for the same camshaft! The point is, the more lifter movement method yields less measured duration, so make sure you compare apples to apples when shopping for a cam. Normally, the 0.050-in. lifter movement is used for advertising high-lift cams—ones that give 0.600-in. or more valve lift. Fortunately duration at 0.050-in. lift is usually supplied on cam cards for checking purposes regardless of lift and advertised duration.

Low Compression Ratios—With lower compression—under 9:1—a "short" duration camshaft is needed. This is because as duration increases, so does overlap; the two cannot be separated. As explained earlier, as overlap increases, so does intake charge exhaust gas dilution, air/fuel charge loss out the exhaust and a resulting cylinder pressure loss at lower engine speeds. So it's better to go with less overlap—duration—so the engine will retain as much cylinder pressure as possible. Cylinder pressure loss with an engine already suffering from low compression regardless of how good the cam is will result in an even soggier performing engine. There's more bad news. Manifold vacuum may suffer so much with a high duration cam that vacuum operated accessories, including power brakes, will not operate satisfactorily. So be conservative when choosing a cam for a street engine, particularly if compression is low. Your best bet may be a factory high-performance cam or equivalent. In terms of engine performance and driveability, remember that it's much better to err on the side of conservatism.

High Compression—For a high compression engine, more duration is fine in terms of lost cylinder pressure, but low manifold vacuum may still be a problem. If not, all you should be concerned with is driveability, economy, legal considerations and power output in the rpm range needed for your application. Above all, don't be swayed by the decals on your hero's car. A pro is paid to put them there and what he actually runs may be different. So for the street, a solid-lifter cam is about as far as you should go. And one from the factory may be the best bet. For open exhaust, off-road use—racing—consider a high lift, short duration cam from an aftermarket cam grinder.

Final Words On Camshafts

Not only have the car manufacturers done a lot of dyno and actual use testing of their cams, so have aftermarket cam grinders. The difference is, originally installed factory street performance cams are milder because of a balance they must maintain between driveability, economy and performance. Meeting emissions standards is something they must both live with. But for the weekend warrior who could care less about economy or driveability, an aftermarket cam grinder will have more to offer. After all, that's his business. And some of these cams are offered in factory motorsports catalogs, so don't ignore the factory's performance equipment. So when choosing a cam, lean heavily on recommendations found in both factory performance catalogs and those from cam grinders. Coupled with the information you've gained in this chapter and recommendations found in catalogs, you should be able to come up with the right cam for your engine. And by all means, DO NOT OVERCAM your engine! ∎

INDEX

A
Angle gauges, 19
Assembly & disassembly tools, 18-20
ATDC (after TDC) ranges, 136

B
BDC (bottom dead center), 155-156, 158-159
    compression strokes, 158
    exhaust strokes, 156
Bearing journals, 54-56
    crank kits, 56
    out-of-round, 55
    regrinding, 54-55
    surface finish, 54
    taper, 55-56
    thrust faces, 55
Bearings
    camshaft, 40-41
    installation, 59
    installing inserts, 118-119
    prepping, 118
    sizing the, 117-119
Blocks
    decks, 52-53
    short, 43-75
Blocks, cleaning, 43-46
    chase threads, 45
    clean coolant passages, 45-46
    commercial cleaning methods, 44
    compressed air, 46
    head gasket checks, 44
    oil galleries, 44
    preventing rust, 46
    scrape gaskets, 45
Blocks, inspection & reconditioning, 46-53
    align-bore or align-hone?, 51
    block decks, 52-53
    checking bore wear, 46-48
    choosing piston ring types, 49-50
    crack testing, 53
    glaze breaking, 49
    main bearing bore, 50-51
    piston-to-bore clearance, 48-49
Blowby defined, 3
Bore tapers, 46
Bore wear, checking, 46-48
    bore tapers, 46
    dial bore, 47
    measuring tapers, 47
    sleeving, 47
BTDC (before TDC), 136
Burettes, 26
Bushing installation, 58-59

C
Calipers, 20-21
Cam, degreeing in, 136-138
    correcting valve timing, 136-138
    finding opening & closing points, 136 Cams

    bearing installers, 110
    worn lobes, 6
Cams, street, 164-165
    high compression, 165
    low compression ratios, 165
Camshafts
    bearings, 40-41
    chain driven, 116-117
    degreeing, 133-141
    degrees, 133
    end play, 116-117
    inspection, 60-61
        cam bearing journals, 61
        measuring lobes, 60-61
    installing the, 115-117
    lobe wear, 15-16
        checking lobe lift, 16
        checking valve lift, 15-16
    plugs, 111
    removal, 37
Camshafts & lifters, 59-61
    cam lobe & lifter terms, 59-60
    camshaft inspection, 60-61
Camshafts, degreeing
    degreeing-in cam, 136-138
    degreeing tools, 134
    finding TDC, 134-135
    rocker arm ratios, 133
Camshafts, selecting, 163-165
    determining one's needs, 164
    street cams, 164-165
Camshafts, theory of, 154-165
    changing valve events, 159-163
    four-stroke cycles, 155-159
    selecting camshafts, 163-165
Caps, numbering, 39
Carbon deposits, 7
Closing ramp, 59
Clutches, 30
Compression
    rings, 127
    strokes, 158-159
    tests, 13
Compression ratio, 103-107
    clearance volume (CV), 103-107
    increasing, 107
    low, 165
    swept volume (SV), 103
Connecting rods, inspecting, 67-69
Core plugs, 41-42, 111-112
Crack testing, 53
    dye penetrate, 53
    Magnafluxing, 53
Crankshafts
    sizing the bearings, 117-119
Crankshafts, assembly preparation of, 119-124
    checking oil clearances, 118-119
    end play, 120-121
    installations, 122-124
    rear main bearing seals, 121-122

    seating thrust bearings, 124
Crankshafts, cleaning & inspection, 53-59
    bearing journals, 54-56
    crankshaft runout, 57
    final cleaning, 58
    journal hardness, 57-58
    pilot bushing or bearing, 58-59
    post-grinding checks, 56-57
CV (clearance volume), 103-107
Cycle, four stroke
    compression strokes, 158-159
    exhaust strokes, 156-157
    ignition, 159
    intake strokes, 157-158
    overlap, 155-156
    power strokes, 156
Cycle, four-stroke, 155-159
Cylinder heads, 33-34, 76-107
    assembly, 97-99
        installed height, 97-98
        retainer clearance, 98
        shim thickness, 97-98
    cleanup & inspection, 79-83
        head warpage, 80
        rocker arm stud replacement, 81-83
    farming out, 76
    head gaskets, 143-144
    installing, 144-145
    milling requirements, 81
    reconditioning guides, 88-90
    teardown, 77-79
        removing valve springs, 77-78
        removing valves, 78
        rocker arm & pivot relationship, 77
        spring shims/seats/rotators, 78-79
        valve removal, 77
    valve guides & stems, 83-90
Cylinder head airflow, 86-88
    improving port flow, 86-87
    manifold to cylinder head matching, 87-88
    modify seats, 87
    modify valves, 87
    stoichiometric, 86

D
Damper installation, 140-141
Degree wheels, 26
Detonation, 7, 10
Diagnosis, 7-10
Dial bore gauges, 24-25
Dial indicators, 23-24
Die grinders, 27-28
Dynamometers, 4, 12

E
End gaps, checking, 124-126
    deburring, 125-126
    filing, 125
Engine assemblies, 108-153
    cam bearings, 113-115

chamfer bearings, 113
crankshaft installations, 117-124
cylinder heads, 143-145
degreeing camshafts, 133-141
exhaust manifolds, 152-153
external bits & pieces, 151-153
fuel pumps, 153
installing plugs, 111-113
installing the camshafts, 115-117
intake manifolds, 149-151
oil pump & pan, 141-143
painting engines, 152
piston & rod installations, 127-131
piston rings, 124-127
preparing the block & bearings, 114-115
supplies, 108-113
    assembly area, 110-111
    gaskets, seals & plugs, 109-110
    lubricants, 108-109
    sealers, 109
    tools, 110
thermostats, 151-152
timing chains & sprockets, 131-138
timing-chain cover & crank damper, 139-141
valve covers, 151
valve-to-piston clearances, 138-139
valvetrains, 145-149
water pumps, 153
Engines, balancing, 70
    internal & external, 70
    process, 70
Exhaust
    manifolds, 31
    plugged, 5
    strokes, 156-157
Eyes, protection of the, 28

F
Feeler gauges, 25
Files, 26-27
Flat-tappet lifters, 59-60
Flexplates, 30-31
Flywheels, 30-31
Fuel delivery, incorrect, 4

G
Gaskets
    head, 14, 33-34
    seals & plugs, 109-110
Gauges
    dial bore, 24-25
    feeler, 25
    ring & feeler, 47-48
    small hole, 23
    snap, 22-23
    telescopic, 22
Guides, reconditioning, 88-90
    integral vs. replaceable guides, 89-90
    knurling, 88
    oversize valve stems, 88-89
    valve guide inserts, 89

H
clutches, 30
drain liquids, 30
engine plates, 31
engine stands or workbenches, 30
exhaust manifolds, 31
flywheels or flexplates, 30-31
Head gaskets, 14, 33-34

blown, 5
head bolts, 144
installation, 143-144
Heads; See also *Cylinder heads*
Holley carburetors, 11
Hydraulic lifters, 7

I
Ignitions, 4, 159
Inertia and compression strokes, 158-159
Inspection tools, 20-26
Intake manifolds, 31-32, 102, 149-151
    blown gaskets, 3-4
    milling requirements, 81
Intake strokes, 157-158

J
Journals
    cam bearing, 61
    hardness, 57-58

L
Leak-down tests, 13-15
    head gaskets, 14
    valves, 14
Lifters, 32-33
    flat-tappet, 59-60
    hydraulic, 7
    roller, 60
Lobe lift, 16
Lobe separation, 133
Lubricants, 108-109
    anti-seize compounds, 109
    Engine Oil Supplement (GM), 109
    moly lube, 109
    Oil Conditioner (Ford), 109
Lubricating systems, 2-4

M
Machinist's blue, 25
Manifolds, intake, 102
    gaskets, 149-150
    installing, 150-151
Micrometers, 21-22
    inside, 22
    outside, 21-22
Mikes, 21-22

N
Noises, 8-10
    connecting rods, 10
    fuel pump, 10
    main bearings, 10
    pinpointing location, 8
    piston rings, 10
    piston slap, 9-10
    valvetrains, 8-9
    wrist pin, 10
Number dies, 20

O
Oil consumption, 2-4
    blown intake manifold gaskets, 3-4
    checking for leakage, 2
    cracked oil passage, 3
    piston rings, 2-3
    valves, guides, & seals, 3
Oil gallery plugs, 42, 112-113
Oil pans, installing, 142-143
Oil pumps, 73-75
    driveshafts, 141-142

end clearance, 75
high volume, 74
installing, 142
modifying, 75
pressure-relief valves, 75
shafts, 142
teardown & inspection, 74-75
Oil rings, 126-127
Oil slinger defined, 132
Opening ramp, 59

P
PCV (positive crankcase ventilation) systems, 2-3
Performance, loss of, 4-7
    engine-related problems, 4-7
    non-engine-related problems, 4
Performance problems, 10-15
    compression tests, 13
    leak-down tests, 13-15
    power balance tests, 12-13
    vacuum tests, 11-12
Performance retainers & keepers, 98
Piston & rods installations
    compressing rings, 129
    installing rod bearings, 128
    positioning crankshaft, 128-129
    preparation, 127-128
    rod side clearance, 131
Piston ring tools
    ring compressor, 19-20
    ring expander, 19
    ring filer, 19
Piston rings
    checking end gaps, 124-126
    choosing types, 49-50
    install rings on pistons, 126-127
    oil consumption, 2-3
Piston/rod disassembly & assembly
    installing pins, 71
    orienting the assembly, 71-72
    removing pins, 71
Pistons, 37-38
    and connecting rods, 61-72
    knurling, 49
Pistons, inspecting, 62-67
    checking for cracks, 65
    inspecting domes, 64
    measuring ring grooves, 66
    oil ring grooves, 66-67
    piston damage, 63
    piston pin bore wear, 64-65
    piston skirt diameter, 63-64
    ring grooves, 65-66
    ring removal, 62-63
Plastigage, 25
Plugs
    camshaft, 111
    core, 41-42, 111-112
    installation of, 111-113
    oil gallery, 42, 112-113
    securing, 112
    Welch, 30
Power balance tests, 12-13
Power strokes, 155-56
Preparation tools, 26-28
Pumps
    fuel, 153
    oil, 73-75
    water, 35, 153
Pushrods, 32, 145-146

R

Ramps
    closing, 59
    opening, 59
Retainers & keepers, performance, 98
Rings
    compression, 127
    compressors, 110
    expanders, 110
    oil, 126-127
Rocker arm stud replacement, 81-83
    converting to screw-in studs, 83
    installation, 82
    oversize studs, 81-82
    removal, 82
    screw-in studs, 83
Rocker arms, 32
    assemblies, 99-102
    individually mounted, 99-100
    and pivot relationship, 77
    shaft mounted, 32
    single mount, 32
Rocker arms, installing, 146-149
    adjustable hydraulic, 147-148
    adjustable mechanical, 148
    non-adjustable hydraulic, 148-149
Rocker arms, shaft-mounted, 100-102
    assembly, 101-102
    disassembly, 100
    measuring clearance, 100-101
Rods, 37-38
Rods, inspecting connecting, 67-69
    bent or twisted, 67
    big end, 68
    inspecting rod bolts, 67-68
    rod bearing bores, 68
    small end, 68-69
Roller lifters, 60
Rope seals, 122
RTV silicone sealers, 109, 121, 149-150, 153

S

Scribes & dividers, 25-26
Sealers, 109, 121, 149-150, 153
Seals, 3
    and plugs, 109-110
    rear main bearing, 121-122
    rope, 122
    slip lip, 121
Seals, removing & installing, 139-140
    dowel pins, 140
    gasket sealers, 140
    prepare damper, 140
Short blocks, 43-75
Slip lip seal, 121
Small hole gauges, 23
Snap gauges, 22-23
Sprockets, 35-37
Steel rules, 26
Steel stamps, 20
Street cams, 164-165
Strokes
    compression, 158-159
    exhaust, 156-157
    intake, 157-158
    power, 156
SV (swept volume), 103

T

Tapers
    bore, 46
    measuring, 47
Taps, 28
TDC (top dead center), 8, 132, 155-159
    cylinders at, 6
    finding, 134-135
    intake strokes, 157
    pistons at, 14
    of power stroke, 155
Telescopic gauges, 22-23
Tests
    compression, 13
    leak-down, 13-15
    power balance, 12-13
    vacuum, 11-12
Three-angle valve seat, 91
Timing cards, reading, 137
Timing chains & sprockets, 35-37, 131-138
Timing sets, 72-73
    chain & sprocket, 73
        double rollers, 73
        gear drives, 73
    removal, 36-37
    worn, 5-6
Timing-chain covers & crank dampers
    damper installation, 140-141
    removing & installing seals, 139-140
TIR (total indicator reading), 57
Tools
    assembly & disassembly, 18-20
    assembly and disassembly
        angle gauges, 19
        number dies, 20
        piston ring tools, 19-20
        steel stamps, 20
        torque wrench, 18-19
    cam bearing installers, 110
    degreeing
        pointers, 134
        positive stops, 134
    engine stands, 110
    ring compressors, 110
    ring expanders, 110
Tools, inspection, 20-26
    burettes, 26
    calipers, 20-21
    degree wheels, 26
    dial bore gauges, 24-25
    dial indicators, 23-24
    feeler gauges, 25
    machinist's blue, 25
    micrometers, 21-22
    plastigage, 25
    scribes & dividers, 25-26
    small hole gauges, 23
    steel rules, 26
    telescopic gauges, 22
    V-blocks, 24
Tools, preparation, 26-28
    die grinders, 27-28
    files, 26-27
    taps, 28
Torque defined, 156
Torque wrench tools
    dial torque wrench, 19
    torquing technique, 18-19

V

Vacuum tests, 11-12
Valve events, changing, 159-163
    exhaust valve openings, 159
    exhaust-valve closings, 160-161
    intake valve openings, 159-160
    intake-valve closings, 161
    piston & crankpin position, 161-162
    valve timing on piston downstroke, 163
    valve timing on piston upstroke, 162-163
Valve grinding
    face angle, 93
    lapping valves, 93
    margin, 92-93
Valve guides, 84-88
    & stems, 83-90
    how much clearance?, 85
    inserts, 89
    measuring valve stems, 85
    valve-guide wear, 84-85
Valve jobs & high mileage engines, 14
Valve lifters, installing, 145
Valve lifts
    checking, 15-16
    curves, 163
Valve seats
    reconditioning, 90-91
    three-angle, 91
Valve springs, 94-97
    removing, 77-78
    terms, 94-95
        free height, 95
        load at installed height, 95
        load at open height, 95
        solid height, 95
        spring rate, 94-95
        squareness, 95
Valve springs, checking, 96-97
    free height, 96
    load, 96-97
    squareness, 96
Valve timing on piston downstroke
    exhaust closes, 163
    exhaust opens, 163
    intake lifts, 163
Valve timing on piston upstroke
    exhaust lift, 162
    intake closed, 162-163
    intake open, 162
Valves
    adjustments, 7
    burned, 6-7
    and carbon deposits, 7
    guides & seals, 3
    and leak-down tests, 14
    reconditioning, 91-94
    removing, 78
Valve-to-piston clearances
    clay method, 138-139
    dial indicator method, 139
Valvetrains
    installing pushrods, 145-146
    installing rocker arms, 146-149
    installing valve lifters, 145
V-blocks, 24

W

Water pumps, 35
Welch plugs, 30
White smoke, 5
Windage trays, 142

ABOUT THE AUTHOR

Tom Monroe built his first engine in 1957 while in high school. It was a flathead Ford, modified with adjustable lifters, special valve springs, "3/4-race" cam, ported, polished and relieved, milled heads and twin Stromberg carburetors and installed in his 1951 Ford. Parts for this project came from retired race engines. More importantly, information came from a three-issue article by Roger Huntington in *Hot Rod Magazine*. This engine building experience lead to bigger things, not to mention rebuilding transmissions and rear axles because of the extra power given the tired flathead. Rounding out his high-school career, Tom crewed on a supermodified dirt-track car race team.

After graduating from college in Mechanical Engineering, Tom took a position with Ford Motor Company as a chassis-design engineer. One of his assignments involved the design, building and testing of LeMans, CanAm and TransAm race cars. During his spare time, he helped campaign a McLaren CanAm road-race car at race courses from Elkart Lake, Wisconsin, to Sebring, Florida.

Later, Tom launched his own engineering-consulting business. Activities included the design, build and testing of Pro Stock drag cars, late-model stock cars, road-race cars, NASCAR modifieds, Formula Super V's and off-road race trucks. In 1981, Tom set the C-Production land-speed record at the Bonneville Salt Flats at a speed of 217.849 mph in 1981. As of this writing the record still stands.

During the mid-70s, Tom was the chief engineer on the Bricklin sports cars. Afterwards he became the automotive editorial director of HPBooks, a position he held until 1987. In addition to doing engineering and race-car consulting work, Tom is instructing in the Engineering and Technologies Division at Catawba Valley Community College in Hickory, North Carolina. To round out his background, Tom is a registered Professional Engineer and ASE certified Master Automotive Technician. ∎

OTHER BOOKS FROM HPBOOKS AUTOMOTIVE

HANDBOOKS
Auto Electrical Handbook: 0-89586-238-7
Auto Upholstery & Interiors: 1-55788-265-7
Brake Handbook: 0-89586-232-8
Car Builder's Handbook: 1-55788-278-9
Street Rodder's Handbook: 0-89586-369-3
Turbo Hydra-matic 350 Handbook: 0-89586-051-1
Welder's Handbook: 1-55788-264-9

BODYWORK & PAINTING
Automotive Detailing: 1-55788-288-6
Automotive Paint Handbook: 1-55788-291-6
Fiberglass & Composite Materials: 1-55788-239-8
Metal Fabricator's Handbook: 0-89586-870-9
Paint & Body Handbook: 1-55788-082-4
Sheet Metal Handbook: 0-89586-757-5

INDUCTION
Holley 4150: 0-89586-047-3
Holley Carburetors, Manifolds & Fuel Injection: 1-55788-052-2
Rochester Carburetors: 0-89586-301-4
Turbochargers: 0-89586-135-6
Weber Carburetors: 0-89586-377-4

PERFORMANCE
Aerodynamics For Racing & Performance Cars: 1-55788-267-3
Baja Bugs & Buggies: 0-89586-186-0
Big-Block Chevy Performance: 1-55788-216-9
Big Block Mopar Performance: 1-55788-302-5
Bracket Racing: 1-55788-266-5
Brake Systems: 1-55788-281-9
Camaro Performance: 1-55788-057-3
Chassis Engineering: 1-55788-055-7
Chevrolet Power: 1-55788-087-5
Ford Windsor Small-Block Performance: 1-55788-323-8
Honda/Acura Performance: 1-55788-324-6
High Performance Hardware: 1-55788-304-1
How to Build Tri-Five Chevy Trucks ('55-'57): 1-55788-285-1
How to Hot Rod Big-Block Chevys: 0-912656-04-2
How to Hot Rod Small-Block Chevys: 0-912656-06-9
How to Hot Rod Small-Block Mopar Engines: 0-89586-479-7
How to Hot Rod VW Engines: 0-912656-03-4
How to Make Your Car Handle: 0-912656-46-8
John Lingenfelter: Modifying Small-Block Chevy: 1-55788-238-X
Mustang 5.0 Projects: 1-55788-275-4
Mustang Performance ('79–'93): 1-55788-193-6
Mustang Performance 2 ('79–'93): 1-55788-202-9
1001 High Performance Tech Tips: 1-55788-199-5
Performance Ignition Systems: 1-55788-306-8
Performance Wheels & Tires: 1-55788-286-X
Race Car Engineering & Mechanics: 1-55788-064-6
Small-Block Chevy Performance: 1-55788-253-3

ENGINE REBUILDING
Engine Builder's Handbook: 1-55788-245-2
Rebuild Air-Cooled VW Engines: 0-89586-225-5
Rebuild Big-Block Chevy Engines: 0-89586-175-5
Rebuild Big-Block Ford Engines: 0-89586-070-8
Rebuild Big-Block Mopar Engines: 1-55788-190-1
Rebuild Ford V-8 Engines: 0-89586-036-8
Rebuild Small-Block Chevy Engines: 1-55788-029-8
Rebuild Small-Block Ford Engines: 0-912656-89-1
Rebuild Small-Block Mopar Engines: 0-89586-128-3

RESTORATION, MAINTENANCE, REPAIR
Camaro Owner's Handbook ('67–'81): 1-55788-301-7
Camaro Restoration Handbook ('67–'81): 0-89586-375-8
Classic Car Restorer's Handbook: 1-55788-194-4
Corvette Weekend Projects ('68–'82): 1-55788-218-5
Mustang Restoration Handbook ('64 1/2–'70): 0-89586-402-9
Mustang Weekend Projects ('64–'67): 1-55788-230-4
Mustang Weekend Projects 2 ('68–'70): 1-55788-256-8
Tri-Five Chevy Owner's ('55–'57): 1-55788-285-1

GENERAL REFERENCE
Auto Math: 1-55788-020-4
Fabulous Funny Cars: 1-55788-069-7
Guide to GM Muscle Cars: 1-55788-003-4
Stock Cars!: 1-55788-308-4

MARINE
Big-Block Chevy Marine Performance: 1-55788-297-5

HPBOOKS ARE AVAILABLE AT BOOK AND SPECIALTY RETAILERS OR TO ORDER CALL: 1-800-788-6262, ext. 1

HPBooks
A division of Penguin Putnam Inc.
375 Hudson Street
New York, NY 10014